Advanced Digital Systems:
Experiments and Concepts with CPLDs

Advanced Digital Systems:
Experiments and Concepts with CPLDs

by
Leo Chartrand

THOMSON
DELMAR LEARNING

Australia Canada Mexico Singapore Spain United Kingdom United States

THOMSON

DELMAR LEARNING

Advanced Digital Systems: Experiments and Concepts with CPLDs
Leo Chartrand

Vice President, Technology and Trades SBU:

Alar Elken

Editorial Director:

Sandy Clark

Senior Acquisitions Editor:

Steve Helba

Development:

Dawn Daugherty

Marketing Director:

Dave Garza

Marketing Coordinator:

Casey Bruno

Production Director:

Mary Ellen Black

Production Manager:

Larry Main

Technology Project Manager:

Kevin Smith

Technology Project Specialist:

Linda Verde

Library of Congress Cataloging-in-Publication Data:
Card Number:

ISBN: 1401866360

NOTICE TO THE READER

Contents

Lab 2 Older Generation Logic Gate Systems Versus CPLD Systems

Lab 3 Flip-Flops, Shift Registers, and Switch Bounce

Lab 4 Serial and Parallel Data Transfer Systems

Lab 5 JK Flip Flop and Counter Fundamentals

Lab 6 Digital Display Decoder System

Lab 7 "1 of X" Decoder and Encoder Systems

Contents

Lab 11 Memory System Fundamentals

Lab 12 Liquid Crystal Displays (LCD)

Appendix A The Evolution of ROM and RAM

Appendix B VHDL Design Guide

Appendix C FLEX Expansion Header Guide

Appendix D Forms and Guides for the DMD Lab 12 Project

Appendix E Summary Sheet for FLEX IC Designs:
Using "Lab 1" as a Guide

Appendix F Altera Simulator Guide

Index

Preface

INTENDED AUDIENCE

Advanced Digital Systems: Experiments and Concepts with CPLDs is a lab manual that is targeted at any college or university that teaches digital principles. It is written at a level that can be easily understood by first-year students. This book includes a set of 12 labs. The labs teach digital fundamental concepts and apply these concepts to complex programmable logic devices (CPLD) lab experiments using the Altera laboratory package. All labs include assignment questions that can be completed outside of the lab and PowerPoint presentations that students can use to review digital concepts. Many labs include design projects that will test a student's ability to apply learned knowledge. Because this book includes an explanation of digital concepts, it can stand on its own as a single source of information. A separate formal textbook is not required to learn the digital principles.

Advanced Digital Systems: Experiments and Concepts with CPLDs can be used at a community college with a 2-year technician program and a 3-year technology program. A 2-year program typically needs "application approach" labs. A 3-year program typically needs "design approach" labs. An "application approach" lab has students building and testing a system studied in class, which has a predetermined response. A "design approach" lab has students building the basic system and then using original thinking to improve or modify the system operation.

To use this book as an "applications approach" lab book, students would build and test the basic system included at the beginning of each lab exercise. The optional design work exercises that follow the basic system can either be omitted or an instructor can provide and study the solution for these design exercises with the students prior to the scheduled lab class. This method essentially converts a "design" lab into an "application" lab. Thus, an instructor controls the "design approach" or "application approach" by the amount of pre-lab solution revealed in class prior to the lab.

To use this book as a "design approach" lab book, students would build and test the basic system included at the beginning of each lab exercise and then follow that up with a selected number of optional design exercises provided in the book. The final result is a book that can be used to force original thinking (design approach), or solutions can be reviewed prior to the lab to convert the lab into an application lab.

Many colleges or universities may find that this book can stand on its own as a single source of digital information. Other colleges or universities that teach a more advanced or intense course in digital principles can use this book in the lab and use a traditional textbook for class work.

A CPLD design minimizes the amount of electrical wiring and eliminates the use of complicated test equipment. As a result of this reduction in engineering complexity, it is feasible for younger, less-experienced students to use this book.

ABOUT THIS BOOK

To understand the value of this book you need to review the history of using digital integrated circuits (ICs) or chips to teach digital fundamentals. Before the advent of the CPLDs, digital fundamentals courses required students to have an electrical principles background. Digital systems were pieced together with single-function transistor transistor logic (TTL) ICs. These designs would often require five or six TTL ICs. The wiring would often be very complex and messy. Digital test equipment would often be needed to make the system functional.

Another problem with TTL IC technology is obsolescence and sourcing laboratory packages. TTL IC technology is over 30 years old and finding a source for replacement parts is more difficult with each passing year. TTL IC technology laboratory packages (also called logic trainer systems) are expensive to replace and difficult to maintain.

Using TTL IC technology prevents students from testing VHDL designs. VHDL is an acronym for the VHSIC Hardware Description Language. VHSIC is an acronym for Very High Speed Integrated Circuit. VHDL is a state-of-the-art hardware design language. Exposing students to this industry-accepted standard is very beneficial. CPLD technology is needed to teach VHDL in a lab environment.

Altera Corporation was founded in 1983. Altera is one of the world pioneer manufacturer of CPLD ICs. The Altera University Program (UP) division was founded to offer schools a CPLD design laboratory package. The package is called the UP board. Using a UP board to build and test digital systems eliminates most of the wiring. A CPLD is a programmable device. Altera also developed the MAX+PLUS® II development system. It is a fully integrated software package that enables design entry, synthesis, verification, design simulation, and CPLD programming. It allows Altera customers to quickly implement designs using CPLDs. A student edition comes with the UP board. A student uses a PC and the MAX+PLUS® II software to draw a diagram of the digital system and then transfers the entire system into one CPLD IC (programs the IC). A single CPLD can easily accommodate a complex digital system design that would require hundreds of older generation TTL ICs.

CPLD technology allows students to concentrate on digital principles and not the electrical wiring. Reducing the wiring complexity also allows instructors to manage large lab group sizes. Helping a student in the lab becomes an exercise in reviewing a digital system diagram as opposed to analyzing messy project boards with four to five ICs with 50 or 60 wires.

The Altera UP board is very affordable ($149 US) compared to TTL-style digital logic trainers costing thousands of dollars. A school can equip an entire 20-station UP board lab at nearly the same cost as a single TTL IC technology laboratory package. Some schools choose to have students purchase their own personal UP board.

Many digital concepts books currently on the market present concepts quickly and all at the same time. Students are shown the entire forest before they get to look at a single tree. Some students find this approach confusing and demoralizing. This book presents concepts incrementally. Students are not asked to digest all the digital principles at once. Care has been taken to ensure students progress systematically from one digital concept to the next. This approach creates a positive class atmosphere because students are confident.

STUDENT CD-ROM

This book includes a CD-ROM with the MAX+PLUS® II software, PowerPoint presentations and the VHDL code for various labs. The students can use the PowerPoint presentations to review digital concepts.

INSTRUCTOR'S CD-ROM

An instructor's CD-ROM is available and it includes many resources:
- The solutions to all lab exercises and lab questions.
- The Altera UP board user guides.

- An instructor's guide that explains how to set up and protect the Altera UP board from damage in the lab.
- Suggestions and guides for lab and classroom activity.
- Tutorial questions. The problem sets can be used in class as a tutorial or they can be handed out as homework assignment questions.
- Solutions to tutorial questions.
- A copy of Appendix E. Appendix E is a summary sheet for FLEX IC designs. It summarizes how to use "Lab 1 as a guide." Extra copies can be printed and left at each station in the lab.
- Lecture note transparencies.
- A discrete IC (TTL IC) lab and lab solution. This lab is optional and it can be used to supplement the CPLD labs.
- PowerPoint presentations.

How to Use This Book

General Guidelines

Each lab contains a list of Equipment, Objectives, Introductory Information, PowerPoint Presentation, Lab Work Procedure, Lab Exercises and Questions, and Conclusions. Most digital fundamental courses are typically scheduled with weekly lecture classes and a weekly lab. An instructor can use the weekly lecture classes to present the material found in the objectives, introductory information, and the PowerPoint presentation. The instructor also can use the weekly lecture class to present and review the Instructor's CD-ROM tutorial questions. This will prepare the students for the weekly lab class and allow them to start immediately at the lab work procedure section of the book when they enter the lab. The instructor can control the complexity of the lab activity. Each lab includes a set of lab exercises that instructors can optionally pick and choose from. Instructors can make a lab more complex by assigning more lab exercises, or they can make it less complicated by reducing the number of lab exercises. Many labs include exercises that can be used as lab projects. Instructors also can control the challenge level by picking and choosing from the various project suggestions. Each lab includes lab questions. These questions are problem sets that can be assigned to students and completed as a homework assignment.

Specific Issues Regarding the First Half and Second Half of the Book

The first half of the book includes Labs 1 to 5. These labs introduce students to basic logic gate theory, shift registers, data registers, and counters. The first five labs of this book are an abbreviation of the material presented in my first book *Digital Fundamentals: Experiments and Concepts with CPLDs*, ISBN: 1401842461, which was published in the fall of 2003. My first book covers the topics presented in Labs 1 to 5 at a slower pace using a series of 16 labs and 4 projects.

The second half of the book includes Labs 6 to 12. These labs take the building blocks presented in the first five labs and apply them to advanced systems using display decoders, multiplexers, keypads, adders, subtractors, ROM, RAM, and dot matrix displays. The last seven labs of this book are presented in great detail at a somewhat slower or relaxed pace.

If you are concerned about the pace used to present the material in the first five labs, there are solutions. The Instructor's CD-ROM that comes with this book includes the resources that accompanied my first book. They include: transparency masters (class notes), all the tutorial worksheets, and tutorial solutions. This means an instructor has access to class notes and worksheets for 16 labs of my first book and can use them to present the information found in the first 5 labs of this book. Another approach would be to use both books. This approach may be better suited to a two-part digital fundamentals course that spans two semesters. Schools that teach a single semester course will likely find that this book, on its own, with the additional resources for Labs 1 to 5 will be sufficient.

ACKNOWLEDGMENTS

The author and Thomson Delmar Learning wish to acknowledge and thank the reviewers for their suggestions and comments during the development of this book. We thank the following:

Ken Reid
Indiana University Perdue University
Indianapolis, IN

Brian Warnecke
DeVry University
Orlando, FL

Steve Lympany
Central Carolina Community College
Sanford, NC

ABOUT THE AUTHOR

Mr. Chartrand holds a Bachelor of Science degree in electrical engineering from Queen's University in Kingston, Ontario. He has been teaching digital courses for 20 years at Niagara College in Welland, Ontario. Mr. Chartrand has made industry contributions with various designs, including interfacing an infrared camera to a PC, creating a digital circuit board used as a PC training system, and designing a control pendant for an air-filled medical bed. He also worked as a plant engineer for General Motors.

DEDICATION

This book is dedicated to my wife Gillian, my son Steffan, and my daughter Claire, who put up with my antics and who support and inspire me unconditionally on a daily basis. I would also like to acknowledge the following family and friends: Ron, Richelle, Joelle, Daniel, Denis, Catherine, David, Zoe, Joan, Gary, Nathan, Faye, Jeff, Jan, Jonathan, Laura, Sarah, Caroline, Thomas, Edith, Joan, Greg, Mike, Liz, and Gerald. A special thanks to Leo Tiberi and my coworkers at Niagara College and to Dawn Daugherty and Dave Garza at Delmar Learning.

Lab 1

Logic Gate Systems

Equipment
Altera UP board
Book CD-ROM
Spool of 24 AWG wire
Wire cutters

Objectives
Upon completion of this lab, you should be able to:

- Use binary numbers to represent digital information.
- Convert numbers into binary.
- Use switches and LEDs with digital systems.
- Operate various logic gates such as NOT, AND, OR, NAND, NOR, XOR, and XNOR.
- Use logic gates to design a vending machine system.
- Use the Altera UP board to test the operation of the vending machine.

INTRODUCTORY INFORMATION

Decimal Number System

The decimal number system is a base 10 number system. It has 10 numerals: 0 through 9. Counting in decimal is easily explained by analyzing a three-digit car odometer. The car odometer is made up of three plastic discs. Each disc is labeled with all the numerals of the number system. The discs spin as the car travels forward. The driver of the car views the odometer by looking at the front of the discs through a rectangular opening. Refer to Figure 1-1.

When the car travels, the right-hand disc (the least significant digit [LSD]) rolls forward 1, 2, 3, 4, 5, 6, 7, 8, and 9. When the LSD has reached the largest decimal numeral, it rolls back to 0 and the middle disc rolls ahead to 1. The driver now sees 010. The LSD cycles to 9 again while the middle disc stays at 1. The driver sees the numbers 10 through 19. The left-hand disc (the most significant digit [MSD]) will move forward after the two other discs wrap around from 9 back to 0. The driver sees 99 change to 100.

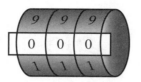 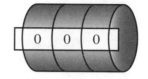

Decimal Odometer   Binary Odometer

Figure 1-1 Decimal and Binary Odometer Limit Resistor.

Binary Numbers

The binary number system is a base 2 number system. It has two numerals: 0 and 1. Counting in binary can be explained by analyzing a 3-binary digit (**bit**) car odometer. The 3-bit odometer is made up of three plastic discs. Each disc is labeled with all the numerals of the number system. In this case the front of the disc shows a 0 and the back of the disc shows a 1. Refer to Figure 1-1. When the car travels, the right-hand disc (the least significant bit [LSB]) rolls forward to 1. The driver sees 001. The LSB has reached the largest binary numeral and it will roll back to 0. The middle disc rolls ahead to 1 and the driver will see 010. The binary number 010 represents 2 in the decimal system. The LSB cycles to 1 again while the middle disc stays at 1. The driver will see 011, which represents 3. The left-hand disc (the most significant bit [MSB]) will move forward only after the two other discs wrap around from 1 back to 0. The driver will see 011 change to 100 (3 changes to 4). This is equivalent to 99 changing to 100 on the decimal odometer.

Table 1-1 shows the complete odometer cycle.

Table 1-1 Binary Odometer Cycle

			Decimal Values
0	0	0	= 0
0	0	1	= 1
0	1	0	= 2
0	1	1	= 3
1	0	0	= 4
1	0	1	= 5
1	1	0	= 6
1	1	1	= 7

Binary Numbers and Digital Systems

Digital systems use binary numbers because transistors can easily represent 0 and 1. A transistor has two operating states called **cut-off** and **saturation**. *Cut-off* is an operating condition equivalent to an everyday light switch in the OFF position. *Saturation* is equivalent to a light switch in the ON position. Transistors can be miniaturized and millions of them placed inside an integrated circuit (IC or chip). How can a simple 1 or 0 (on or off) represent useful information? The answer is simple: A single 1 or 0 is not useful; however, a group of many bits working together is. Later, you will see how a group of 4 bits is used to represent the operation of a vending machine system.

Decimal Positional Weight Chart (PWC)

To convert a binary number to decimal you need to understand the decimal PWC. Figure 1-2 shows a decimal PWC with the number 873 placed inside of it. There are names for each weighted position in the chart.

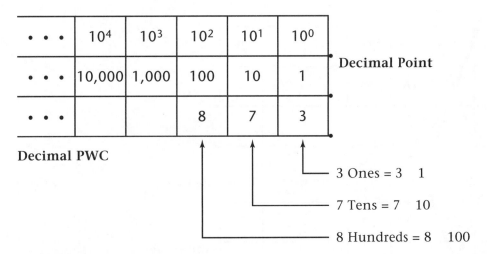

Figure 1-2 Decimal PWC with the Number 873

Use the Binary Positional Weight Chart (PWC) to Convert Binary Numbers to Decimal

Changing the base of the number from 10 to 2 creates the binary PWC. Figure 1-3 shows a binary PWC with the binary number 110 placed inside of it. Figure 1-3 also shows how the PWC is used to convert binary 110 to the decimal number 6.

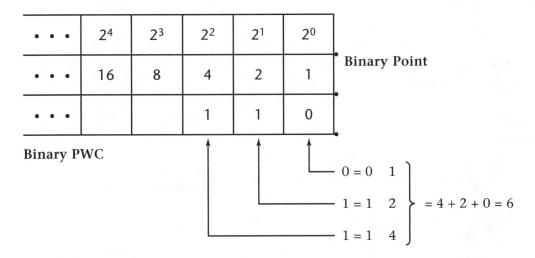

Figure 1-3 Binary PWC with the Binary Number 110

Convert Decimal Numbers to Binary

The PWC can also be used to convert decimal numbers to binary. Figure 1-4 shows how to convert the decimal number 25 to binary. Begin with a binary PWC with empty boxes below it. The number of boxes in the PWC should not extend beyond the total weight of the number. Start at the MSB with a running total of 0. Place a 1 in the empty box if the weight of the current position keeps the running total under or equal to 25. Place a 0 in the empty box if the weight of the current position makes the running total exceed 25. Continue to work toward the LSB. The completed PWC is the number 25 in binary.

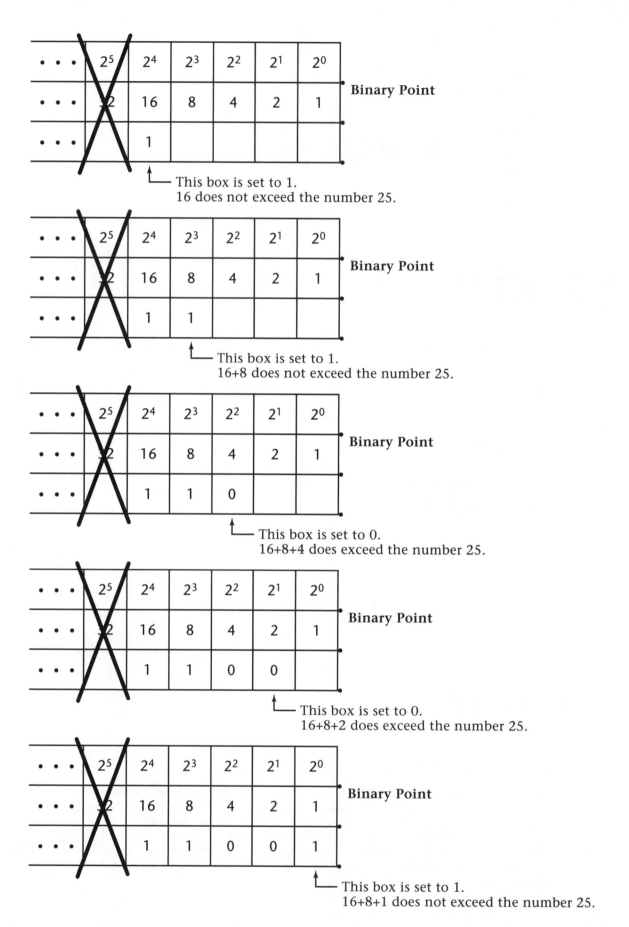

Figure 1-4 Use PWC to convert 25 to binary.

Switches

Students can create and test digital systems by using switches to represent binary input data. A light switch on a wall has an *on* position and an *off* position. A switch used to test digital systems will have a **logic 0** position and a **logic 1** position. Switches are connected to power sources to generate logic 1 and logic 0 signals. To simplify the theory, a 5-volt (V) battery is used to represent the power source. Refer to Figure 1-5. Connect the switch to the positive, "+," side of the battery to represent logic 1. Connect the switch to the negative, "−," side of the battery to represent logic 0. The negative side is also called ground or Gnd or 0 volt. Many digital systems use 5 volts to represent logic 1 and Gnd to represent logic 0. A schematic diagram uses an arrow to 5 volts and a ground (Gnd) symbol to replace the terminals of the battery.

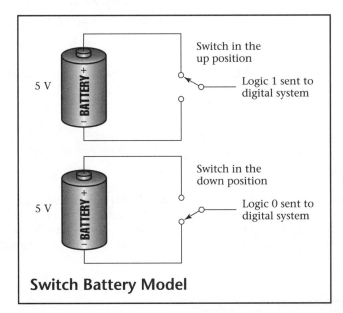 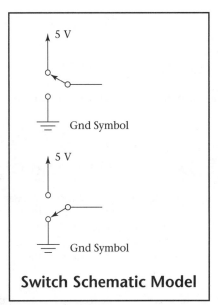

Figure 1-5 Switch Battery Model and Switch Schematic Symbol

LEDs

Students can create and test digital systems by using light-emitting diodes (LEDs) to represent binary output data. Output data is the response of the digital system to the input conditions. It is what comes out of your system. There are two ways to connect LEDs to a digital system. One method is called **active high** (Figure 1-6[A]) and the other method is **active low** (Figure 1-6[B]). If the digital system responds with a "logic 1" at the output, then the active high LED will light. If the digital system responds with"logic 0," then the active high LED will *not* light. An active high LED is intuitively easy to understand, **High = Light ON**. It is used in upcoming system diagrams. An active low LED, on the other hand, is an LED in reverse. It lights when the system response is "logic 0" and turns off when the system response is "logic 1." This backward response is not as intuitive and takes time to get used to. Figure 1-6(C) shows a resistor placed in series with the LED. It is required to limit the current through the LED. Most often the resistor is omitted in order to simplify an LED diagram. Also, many integrated circuits automatically limit the current and do not need the resistor.

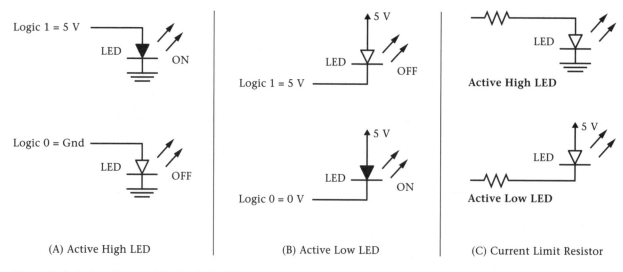

Figure 1-6 Active High and Active Low LEDs

Switches, LEDs, and Digital Systems

To further understand how switches and LEDs are used, let's study a digital system that adds binary numbers. A digital system that adds two 2-bit numbers would require four switches to represent the inputs and three LEDs to represent the outputs. The smallest sum is zero [S = 0 + 0]. The largest sum is 6 [S = 3 + 3 (3 = 11 in binary)]. Three LEDs are required to display the largest sum (6 = 110 in binary). Refer to Figure 1-7.

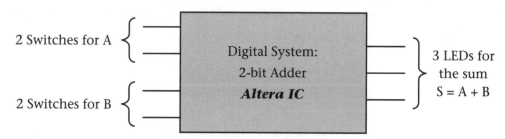

Figure 1-7 Digital System: 2-Bit Adder

The NOT Gate

The NOT gate outputs the inverse of the logic level applied to the input. The **NOT** gate is also called the **inverter** gate. The truth table defines the operation of the gate. The NOT gate can be tested with one switch and one LED. Use the switch to place all possible values at the input while monitoring the LED to observe the gate's output response. Refer to Figure 1-8.

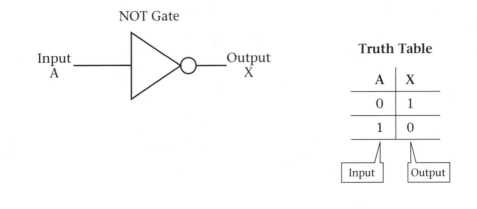

Truth Table

A	X
0	1
1	0

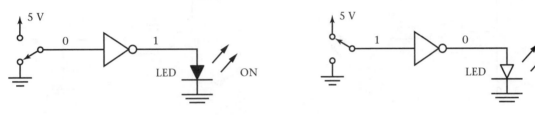

Figure 1-8 The NOT Gate

The AND Gate

The AND gate outputs a logic 1 if inputs A and B are both 1. The truth table defines the operation of the gate. The AND gate can be tested with two switches and one LED. Use the switches to place all possible values at the input while monitoring the LED to observe the gate's output response. Refer to Figure 1-9.

The OR Gate

The OR gate outputs a logic 1 if any input, A or B, is 1. The truth table defines the operation of the gate. Refer to Figure 1-10.

Truth Table

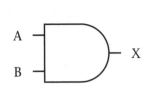

A	B	X
0	0	0
0	1	0
1	0	0
1	1	1

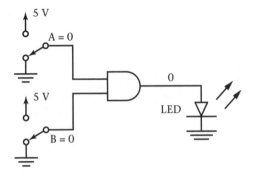

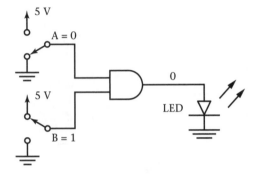

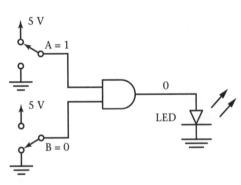

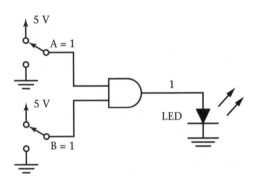

Figure 1-9 The AND Gate

Truth Table

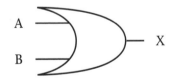

A	B	X
0	0	0
0	1	1
1	0	1
1	1	1

Figure 1-10 The OR Gate

The Three-Input AND Gate

Figure 1-11 shows a diagram of a three-input AND gate. The operation of this three-input AND gate can be easily defined using a truth table. Generating a three-input truth table is a little more complicated than generating a two-input truth table. The PWC technique simplifies the task.

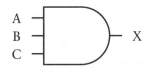

Figure 1-11 The Three-Input AND Gate

PWC Technique for Generating a Truth Table

Step 1: Determine the size (number of rows) of the truth table. Each input can have two values, 0 or 1 (binary), and we have three inputs.

rows = 2^N where N is the number of inputs

Thus, # rows = 2^3 = 8 rows

Refer to Figure 1-12(A).

Step 2: Fill in the columns. Place a binary PWC under the last row of the truth table. The PWC numbers indicate how many groups of alternating 1s and 0s must be placed in each column. Refer to Figure 1-12(A). Fill in each column. Refer to Figure 1-12(B). The AND gate only outputs a logic 1 when inputs A, B, and C are all 1. Refer to Figure 1-13.

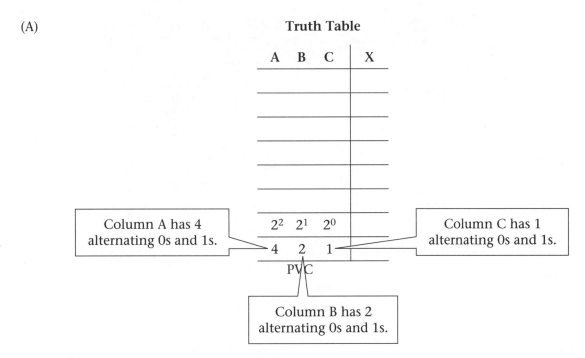

Figure 1-12(A) Use the PWC to lay out the truth table.

(B)

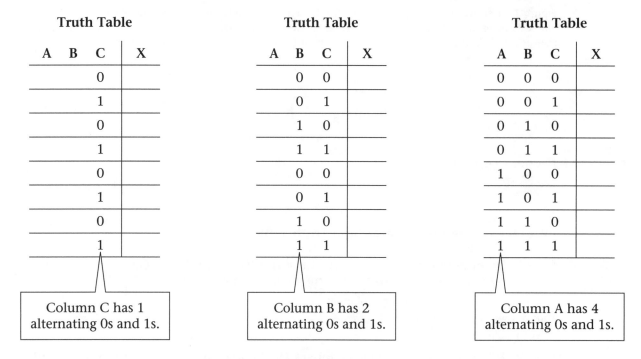

Truth Table

A	B	C	X
		0	
		1	
		0	
		1	
		0	
		1	
		0	
		1	

Column C has 1 alternating 0s and 1s.

Truth Table

A	B	C	X
	0	0	
	0	1	
	1	0	
	1	1	
	0	0	
	0	1	
	1	0	
	1	1	

Column B has 2 alternating 0s and 1s.

Truth Table

A	B	C	X
0	0	0	
0	0	1	
0	1	0	
0	1	1	
1	0	0	
1	0	1	
1	1	0	
1	1	1	

Column A has 4 alternating 0s and 1s.

Figure 1-12(B) Use the PWC to lay out the truth table.

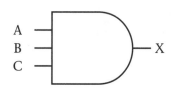

Truth Table

A	B	C	X
0	0	0	0
0	0	1	0
0	1	0	0
0	1	1	0
1	0	0	0
1	0	1	0
1	1	0	0
1	1	1	1

Figure 1-13 Three-Input AND Gate Truth Table

The Three-Input OR Gate

The OR gate outputs a logic 1 if any input, A or B or C, is 1. Refer to Figure 1-14.

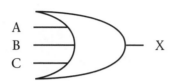

Truth Table

A	B	C	X
0	0	0	0
0	0	1	1
0	1	0	1
0	1	1	1
1	0	0	1
1	0	1	1
1	1	0	1
1	1	1	1

Figure 1-14 Three-Input OR Gate Truth Table

The NAND Gate and the NOR Gate

A *NOT* gate and an *AND* gate combine to create a **NAND** gate. The NAND gate outputs a logic 0 if inputs A and B are both 1. The truth table defines the operation of the gate. A *NOT* gate and an *OR* gate combine to create a **NOR** gate. The NOR gate outputs a logic 0 if any input, A or B, is 1. The truth table defines the operation of the gate. Refer to Figure 1-15.

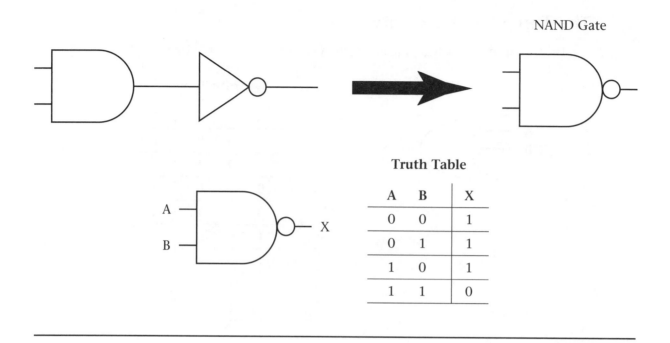

NAND Gate

Truth Table

A	B	X
0	0	1
0	1	1
1	0	1
1	1	0

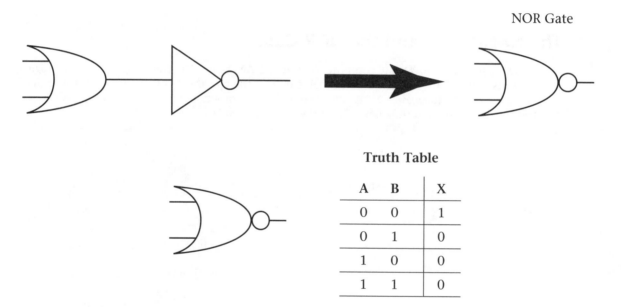

NOR Gate

Truth Table

A	B	X
0	0	1
0	1	0
1	0	0
1	1	0

Figure 1-15 The NAND and NOR Gates

The XOR Gate and the XNOR Gate

The XOR gate is also called the **exclusive OR** gate. The XOR gate outputs a 1 if there is an exclusive logic 1 at the inputs. "Exclusive" means only one input can be high. The XNOR is the exclusive NOT OR gate. The truth table defines the operation of the gates. Refer to Figure 1-16. The XNOR can be used to compare binary data. When A = B, then X = 1.

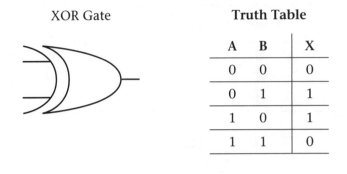

XOR Gate   Truth Table

A	B	X
0	0	0
0	1	1
1	0	1
1	1	0

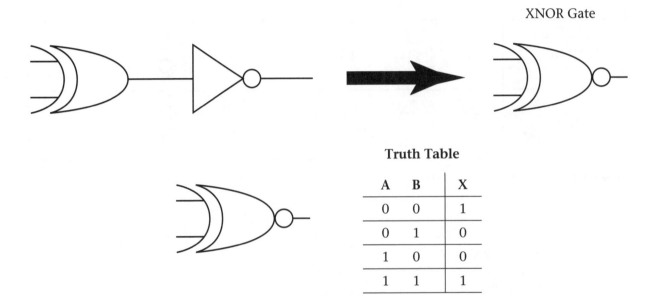

XNOR Gate

Truth Table

A	B	X
0	0	1
0	1	0
1	0	0
1	1	1

Figure 1-16 The XOR and the XNOR Gates

Alternate Logic Gate Symbols

The complete list of logic gates include NOT, AND, OR, NAND, NOR, XOR, and XNOR. The NOT, AND, OR, NAND, and NOR have alternate logic gate symbols. You may encounter alternate symbols so it is a good idea to know what they are. Figure 1-17 shows the alternate symbols.

The operation of an alternate symbol can be explained using a truth table and remembering the difference between active high and active low inputs and outputs. An active low input or output terminates with a bubble (or circle). An active high input or output terminates *without* a bubble. For example, study the standard NAND gate and the alternate NAND gate. Refer to Figure 1-18.

Name of Gate	Standard Symbol	Alternate Symbol
NOT	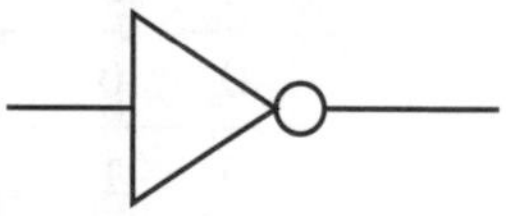	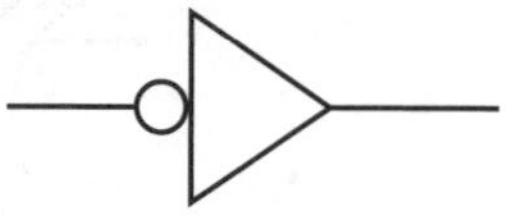
AND	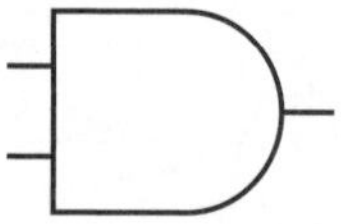	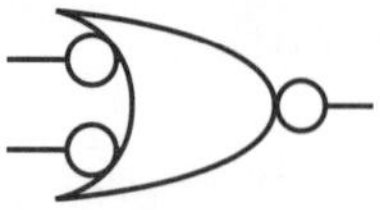
OR	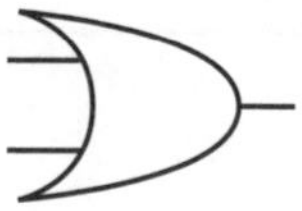	
NAND	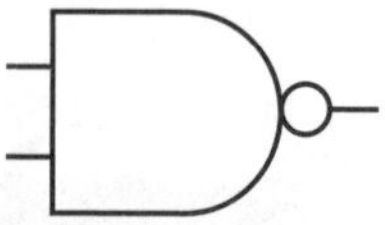	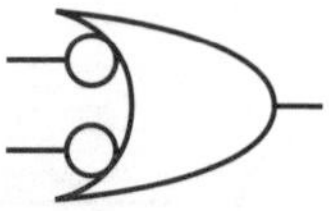
NOR	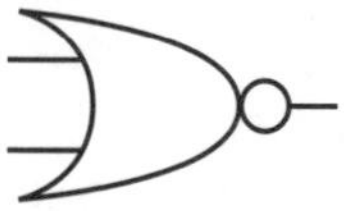	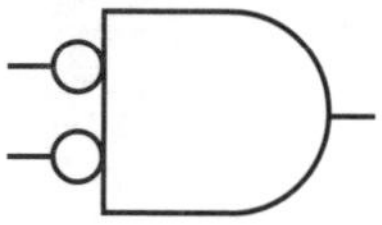

Figure 1-17 Standard and Alternate Logic Gate Symbols

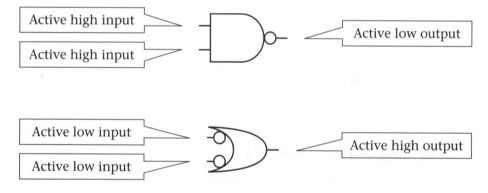

Figure 1-18 Active High and Active Low Inputs and Outputs

An active high input or output is asserted with a 1 (5 volt = high voltage). An active low input or output is asserted with a 0 (0 volt = low voltage). Read the descriptive statements for each gate and immediately study the truth table. Refer to Figure 1-19.

Both statements describe the same NAND gate truth table. This means both symbols are equivalent to each other.

You can derive the alternate symbol for a gate by remembering these two steps:

1. Change the body shape: OR becomes AND; AND becomes OR.
2. Change the input/output termination. Remove existing bubbles and add missing bubbles.

Figure 1-20 shows how to convert the standard NAND symbol to its alternate symbol.

This conversion technique can also be used to figure out the name of the alternate symbol. For example, study the alternate symbol for the NAND gate in Figure 1-20. Apply the conversion procedure to the alternate gate—change the body shape; change the termination (remove and

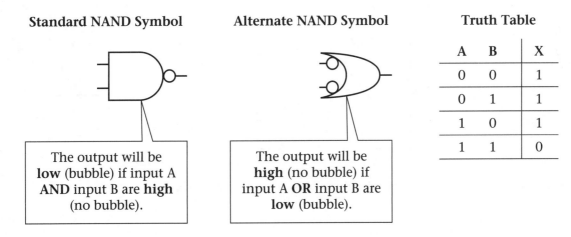

A	B	X
0	0	1
0	1	1
1	0	1
1	1	0

Standard NAND Symbol

The output will be **low** (bubble) if input A **AND** input B are **high** (no bubble).

Alternate NAND Symbol

The output will be **high** (no bubble) if input A **OR** input B are **low** (bubble).

Truth Table

Figure 1-19 Standard and Alternate NAND Gates

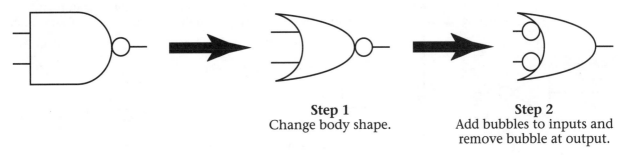

Step 1
Change body shape.

Step 2
Add bubbles to inputs and remove bubble at output.

Figure 1-20 Deriving the Alternate Symbol for a NAND Gate

add bubbles). The alternate symbol reverts back to the standard symbol. This single procedure can be used to convert standard symbols to alternate symbols and to revert alternate symbols back to standard symbols in order that they be identified.

When Are the Alternate Symbols Used?

As we have seen most gates have two symbols to depict their operation. One symbol is the **standard** symbol and the other is the **alternate** symbol. When designing a digital system you must consider the factors that determine when to use a standard symbol or when to use an alternate symbol. There is a general rule of thumb that determines which symbol should be used. A look into active high and active low devices is necessary to understand the rule. Motion sensors, door and window intrusion sensors, keyboard switches, LEDs, and digital displays are all examples of devices that can be connected to logic gate systems. Some devices are active high and some are active low. If you review what you know about LEDs, you will remember that an active high LED lights with a "logic 1," and an active low LED lights with a "logic 0." This is the factor that determines whether the standard or the alternate symbol is used.

Here is the rule:

Connect an active low gate (terminates with a bubble) to an active low device.
Connect an active high gate (does *not* terminate with a bubble) to an active high device.

NOTE: The rule is a general guide, and some system designers choose not to use it.

EXAMPLE: *Connect each type of an LED to a NAND gate using the rule.*

An LED can be connected as active high or as active low. The NAND gate can be drawn using the standard symbol or the alternate symbol. According to the rule, the active high LED would be connected to the alternate NAND symbol, and the active low LED would be connected to the standard NAND symbol. Refer to Figure 1-21.

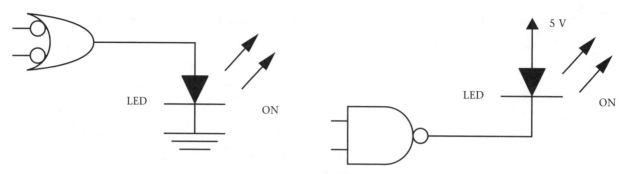

Figure 1-21 Connect an LED to a standard and an alternate NAND gate symbol.

Logic Gate Equations

Each logic gate has a **Boolean** equation to represent its operation. Boolean equations use math operators (+, x) to represent the function of the logic gate. Boolean math shares operators and some operating rules with regular math but that is where the similarity ends. Refer to Figure 1-22.

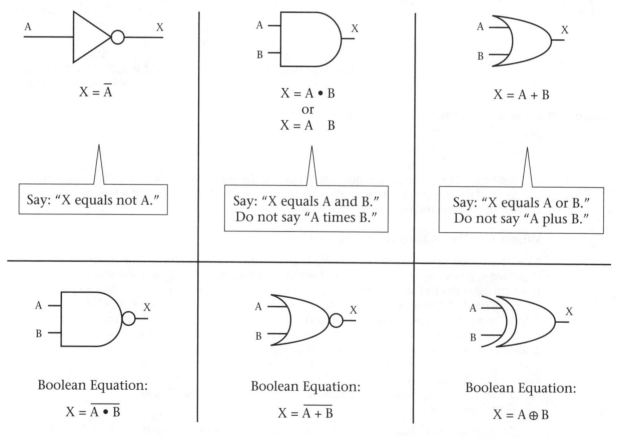

Figure 1-22 Logic Gate Boolean Equations

Vending Machine System

Logic gates will be combined to create a digital system for a vending machine. A logic gate equation will be used to represent the operation of the vending machine system. The vending machine will be placed at the entrance of a mall. The machine will dispense a 75-cent plastic egg with a toy inside it. The machine will only accept quarters (25 cent coins) or a dollar ($1 coin in Canada or a $1 note in the United States). To simplify the design, the system used to detect and identify the denomination of the currency will not be part of the design.

Step 1: Declare I/O.

Input Variables:

Logic 1 = money inserted. Logic 0 = money *not* inserted.

Quarters: Q1= first quarter, Q2 = second quarter, and Q3= third quarter.

Dollar: L

Output Variables:

Logic 1 = dispense item. Logic 0 = do NOT dispense item.

Toy Package: P

Quarter Change: C

Step 2: Write the equation for the system.

P = Q1·Q2·Q3 + L (Package = Quarter1 and Quarter2 and Quarter3 ... OR ... Dollar)

C = L

> **NOTE:** *Q1·Q2·Q3 is called a product term (P-Term). The variables are separated by the Boolean equivalent of the multiplication symbol, and product means multiply. A P-Term can be treated as a single Boolean equation entity or a single block.*

Step 3: Draw the system diagram. Draw lines to represent the inputs on the left side of the page and the outputs on the right. Refer to Figure 1-23(A).

Work from output P backwards toward the input. Try to figure out the logic gate connected to P. Remember, you can group the P-Term as a single block.

P = Q1·Q2·Q3 + L becomes P = []+ L

Thus, the gate connected to P is OR. Refer to Figure 1-23(B).

Now work from the OR gate backward toward the input. L is a direct connection. The [] symbol is Q1·Q2·Q3, which is a three-input AND gate. Refer to Figure 1-23(C).

C = L connect the wire back from C to L. Refer to Figure 1-23(D).

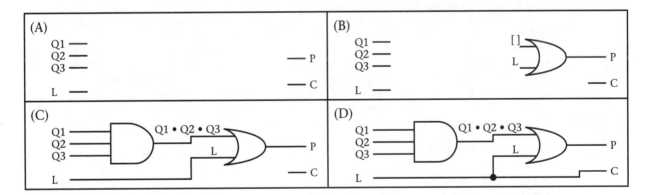

Figure 1-23 Vending Machine System Diagram

Figure 1-24 shows how the digital system responds to the coin sensors.

Altera UP Board

You have seen how logic gates are combined to create digital systems. You need to learn how the Altera UP board is used to test a digital system design. Altera is the name of a corporation that is one of the world's leading pioneers of system-on-a-programmable-chip (SOPC) solutions. Programmable logic devices (PLDs) are semiconductor chips that can be customized and

3 Quarters Inserted

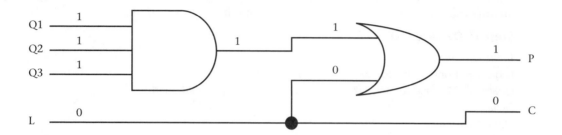

Dollar Inserted

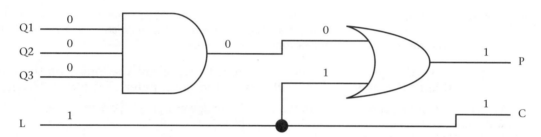

Figure 1-24 Vending Machine System Response

programmed by Altera customers using software tools that run on personal computers. PLDs help Altera customers quickly bring their products to market.

Altera was founded in 1983 by Robert Hartman, Michael Magranet, Paul Newhagen, and Jim Sansbury. The Altera team developed the first reprogrammable logic devices. The first commercialized Altera PLDs, the Classic™ devices, are still sold to Altera customers today. These smaller devices have since been categorized as **SPLDs**, where *S* represents the word **SIMPLE**. Altera expanded its technology leadership in 1988 with the product-term-based **MAX®** architecture and, in 1992, with the look-up table (LUT)-based **FLEX®** architecture. MAX and FLEX IC technology is more advanced and is categorized as **CPLDs**, *COMPLEX PLDs*. Today, Altera sells a wide range of product families, with a total of more than 550 device and packaging combinations.

The Altera **University Program (UP)** division was founded to offer schools a design laboratory package. Two versions of the package are currently in existence: the **UP-1** board and the **UP-2** board. Both boards include the **MAX EPM7128S** device, which has a capacity of 2,500 gates. Gate capacity is a measure of how many logic devices can be programmed into a single Altera IC. Older IC technology contained about 3 logic gates per IC, thus, a capacity of 2,500 gates would mean that a single MAX IC could replace 800 older ICs. There are other factors that fit into this calculation nonetheless; this gives a rough estimate of the benefit of Altera ICs. The UP-1 includes the **FLEX EPF10K20** device, which has a capacity of 20,000 gates. The UP-2 includes the **FLEX EPF10K70**, which has a capacity of 70,000 gates. Refer to Figure 1-25.

> **NOTE:** *Altera is a trademark and service mark of Altera Corporation in the United States and other countries. Altera products are the intellectual property of Altera Corporation and are protected by copyright laws and one or more U.S. and foreign patents and patent applications.*

Altera created the **MAX+PLUS®** II development system. It is a fully integrated software package that enables design entry, synthesis, verification, and device programming. In simpler terms it means: You draw a diagram that represents your digital system, you test and verify the integrity of your system using a simulator, and you transfer your design into an Altera IC. A **simulator** is a program that allows you to test a design before it is programmed into an Altera IC.

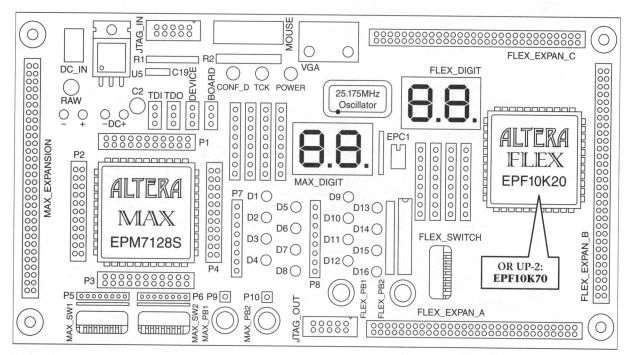

Figure 1-25 Altera UP Board

A simulation can help find errors. Appendix F describes how to use the Altera simulator. This process allows Altera customers to quickly implement designs using CPLDs. A student edition comes with the UP board. This lab book refers to the **UP-1** and the **UP-2** boards as the **UP** board. Although this book focuses on Altera because of the UP board, there are other companies that make PLDs. An example of such a company is **XILINX** Inc.

Altera UP Board Switches and LEDs

The **ALTERA UP Education Board** allows you to build and test digital systems. Switches are available to send binary data to your system. LEDs are available to display the binary response of your system. Two cables must be connected to the UP board: the power-pack cable and the data cable. The data cable is called the Byte Blaster (BB) cable. The BB cable links the Altera integrated circuits (ICs) or chips to the PC's via the printer port (LPT1). Refer to Figure 1-26.

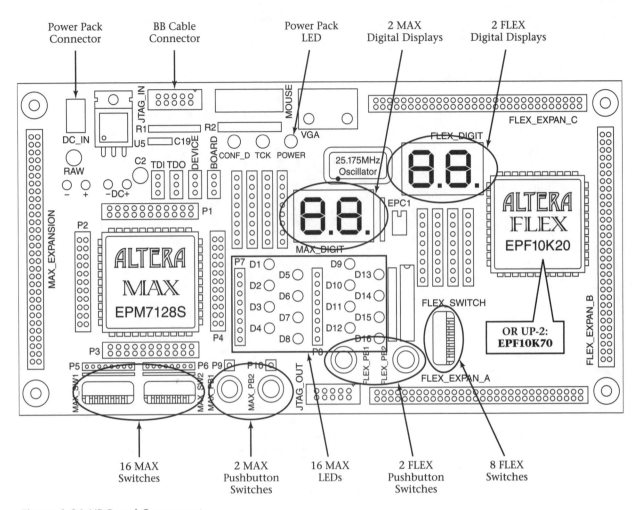

Figure 1-26 UP Board Components

MAX Designs

A digital system design can be programmed and tested using the MAX or the FLEX IC. A MAX design requires wiring to connect the MAX switches and MAX LEDs to the MAX IC. Wiring is accomplished by inserting wires into female header sockets. Female header sockets surround the MAX IC and allow you to insert wires to make connections to the IC. The connection points on the IC are called IC pins. The pins allow transfer of binary data into and out of the MAX IC. The MAX IC has 84 pins. Female header sockets are also attached to the MAX LEDs and MAX switches.

Figure 1-27 shows the MAX LEDs and MAX switches.

LED Positions:

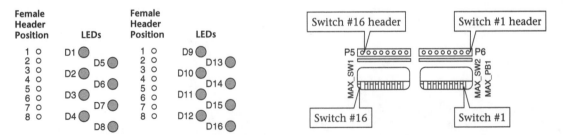

Figure 1-27 UP Board MAX LEDs and MAX Switches

LEDs D1 through D8 are connected in the same sequence to the female headers (i.e., D1 is connected to position 1, and D2 is connected to position 2). LEDs D9 through D16 are connected in the same sequence to the female headers (i.e., D9 is connected to position 1, and D10 is connected to position 2). UP board LEDs are active low. Placing NOT gates on your system diagram can be used to revert them back to active high.

Altera switches are designed to output logic 1 in the up position and a logic 0 in the down position. The female header layout for the switches is simple. Each pin on the header is attached to the switch adjacent to it.

Figure 1-28 shows how wires are used to connect the switches and LEDs to the MAX IC.

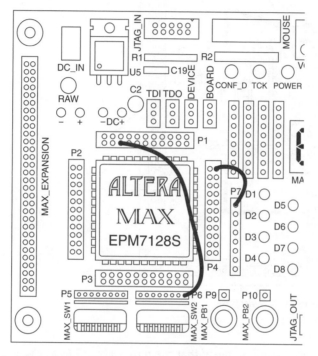

Figure 1-28 Connecting the UP Board MAX Switches and LEDs

The MAX digital displays and the MAX pushbuttons are explained in an upcoming lab.

FLEX Designs

A digital system design can be programmed and tested using the MAX or the FLEX IC. A FLEX design requires no wiring to connect the FLEX switches and LEDs to the FLEX IC. The Altera software is used to assign LED and switch connections. Printed circuit board wire connections (PCB traces) connect the switches and LEDs to the FLEX IC. The FLEX IC has 240 connection points across 4 sides of the IC (60 per side). These connection points are called pins and they transfer binary data into and out of the FLEX IC. The FLEX IC does not have individual round-type LEDs; however, the digital display can be used to create two banks of four bar-type LEDs. Figure 1-29 shows the FLEX LEDs and switches.

FLEX LEDs are active low. Placing NOT gates on your system diagram can be used to revert them back to active high. The FLEX pushbuttons are explained in an upcoming lab.

Older Generation ICs

Before the advent of Altera ICs, digital systems were pieced together with single-function ICs. Two ICs would be required to build the vending machine system. A three-input AND gate IC would be connected to a two-input OR gate IC. More complex designs would often require dozens (at times hundreds) of single-purpose ICs to build a digital system. An example of an older generation IC technology is **TTL ICs** (transistor, transistor, logic). A TTL design often required students to interconnect 5 to 15 ICs. The wiring would often be very complex and

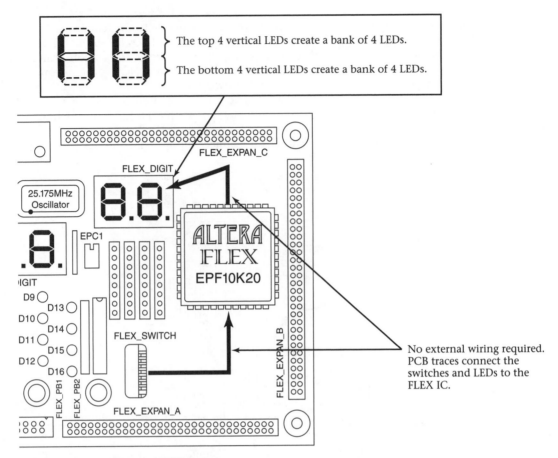

Figure 1-29 UP Board FLEX LEDs and FLEX Switches

messy. Complex designs looked so messy that the term *spaghetti factory* was used to describe the appearance of the wiring. Figure 2-5 in the next lab shows an example of a TTL design and all of the messy wiring. The new Altera **MAX** and **FLEX** ICs can easily accommodate a complex digital system design that would require hundreds of older generation TTL ICs. To build and test the operation of the vending machine system using Altera ICs requires little wiring. The wiring between logic gates is internally programmed into the Altera IC and is not required. All you need to do is connect the input switches and the output LEDs to the Altera IC. Design changes or revisions are also very simple. Older generation IC technology would require a complete teardown and rebuild when changes were required. Altera ICs are easily reprogrammed.

POWERPOINT PRESENTATION

Use PowerPoint to view the Lab 01 slide presentation.

LAB WORK PROCEDURE

Design a Vending Machine System

1. Turn on the computer (PC).
2. Insert a blank formatted floppy into drive **A**. Use Windows Explorer to quick format the floppy. This will erase all files. Use Windows Explorer to create a folder (directory) on the floppy called **Labs**. The project files will be saved in this folder.
3. Start the **MAX+PLUS II** program. This will begin a MAX+PLUS II session. You will see the MAX+PLUS II Manager window. Refer to Figure 1-30.

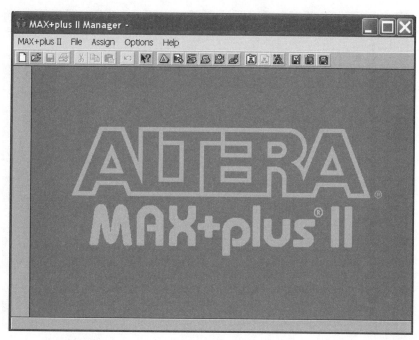

Figure 1-30 Altera MAX+PLUS II Manager

Create a project file: Click the **File** menu and select **New**. You will see the "New" window. Refer to Figure 1-31. Select **Graphic Editor File** and click the "OK" button. You will see the Graphic Editor window inside the MAX+PLUS II window. Refer to Figure 1-32. Save the project. Click the **File** menu, select **Project** and then **Set Project To Current File**. You will see the "Save As" window. Refer to Figure 1-33.

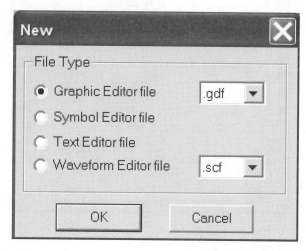

Figure 1-31 Altera "New" Window

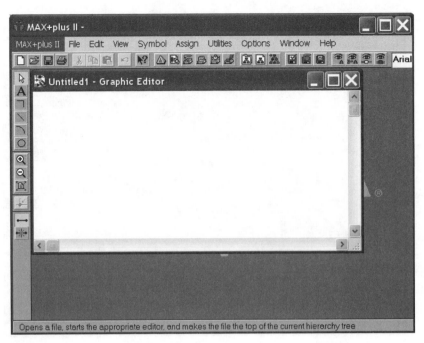

Figure 1-32 Altera "Graphic Editor" Window

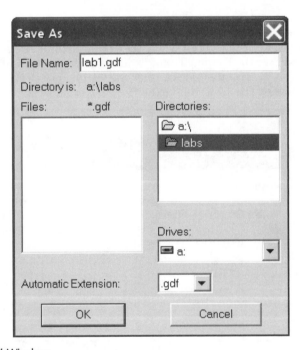

Figure 1-33 Altera "Save As" Window

Replace the word **Untitled2.gdf** with **Lab1.gdf.** Select A: from the **DRIVES:** drop down menu and then select directory (folder) **Labs** (double-click on the folder name). Click the "OK" button. The titles at the top of the Graphic Editor window and the MAX+PLUS II window will show the title of the project: **Lab1.**

4. Enter the logic gate symbol for the OR gate. With the selection pointer (pointer you can move with the mouse), double-click in an empty space in the Graphic Editor window. You will see the "Enter Symbol" window. Refer to Figure 1-34.

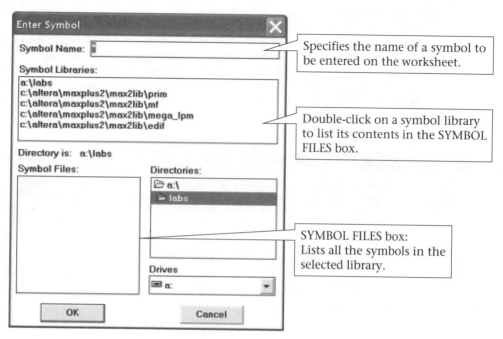

Figure 1-34 Altera "Enter Symbol" Window

Double-click on the **c:\maxplus2\max2lib\prim** library. This will display the symbols in the "primitive" library in the "Symbol Files" box. Locate and double-click on the **OR2** symbol from the menu. The symbol will be placed on the Graphic Editor worksheet.

5. Repeat the symbol entry procedure (step 4) to enter the logic gate symbol for the AND gate (symbol "AND3") and two "NOT" gates. The Graphic Editor window should resemble Figure 1-35.

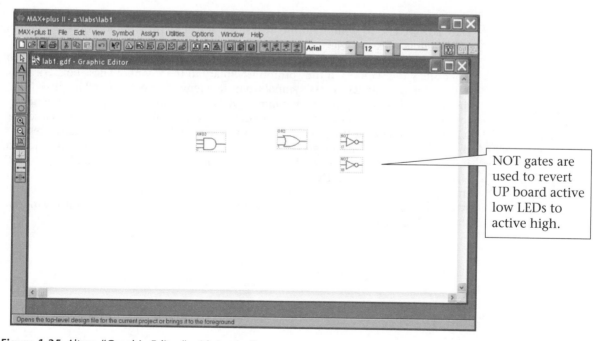

Figure 1-35 Altera "Graphic Editor" with Logic Gates

6. Place an **INPUT** symbol on the worksheet. It will identify the location of an input switch. Double-click the selection pointer to the left of the "AND" gate to open the "Enter Symbol" window. Double-click on the **c:\maxplus2\max2lib\prim** library. This will display the symbols in the "primitive" library in the "Symbol Files" box. Locate and double-click the **INPUT** symbol from the menu. The symbol will be placed on the Graphic Editor worksheet. Repeat this process three more times to enter "INPUT" symbols for the AND gate and the OR gate. The Graphic Editor window should resemble Figure 1-36.

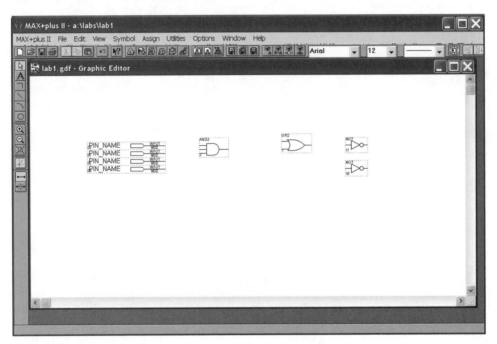

Figure 1-36 Altera "Graphic Editor" with Logic Gates and Input Symbols

7. Place an "**OUTPUT**" symbol on the worksheet. It will identify the location of an output LED. Double-click the selection pointer to the right of the "NOT" gates to open the "Enter Symbol" window. Double-click on the **c:\maxplus2\max2lib\prim** library. This will display the symbols in the "primitive" library in the "Symbol Files" box. Locate and double-click the **OUTPUT** symbol from the menu. The symbol will be placed on the Graphic Editor worksheet. One more "OUTPUT" symbol is required. This time, try to copy the output symbol instead of repeating the symbol entry procedure. Place the selection pointer over the "OUTPUT" symbol. Press and hold down the <ctrl> key on the keyboard while clicking and dragging the mouse. Release the mouse button to drop the symbol on the worksheet. The Graphic Editor window should resemble Figure 1-37.

8. Name the input symbols. Double-click the selection pointer on the word **PIN_NAME** of the top input symbol. Type in **Quarter1**. Repeat the procedure on the other three input symbols. Use the names **Quarter2**, **Quarter3**, and **Dollar.** The Graphic Editor window should resemble Figure 1-38.

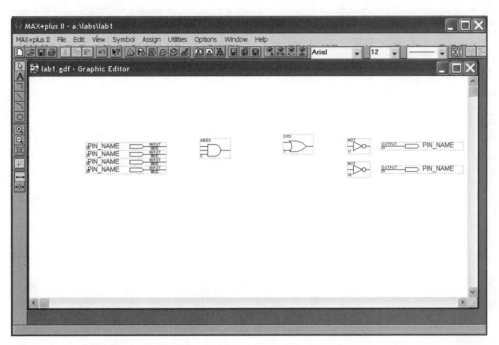

Figure 1-37 Altera "Graphic Editor" with Logic Gates, Input/Output Symbols

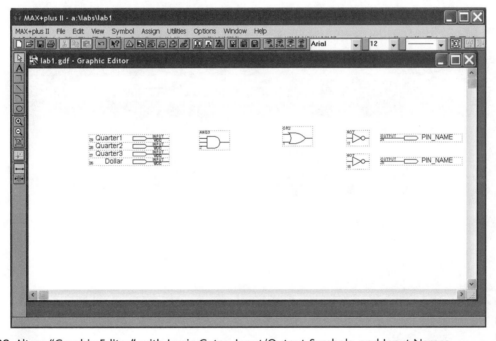

Figure 1-38 Altera "Graphic Editor" with Logic Gates, Input/Output Symbols, and Input Names.

9. Name the output symbols. Double-click the selection pointer on the word **PIN_NAME** of the top output symbol. Type in **Package**. Repeat the procedure on the other output symbol. Use the name **Change**. The Graphic Editor window should resemble Figure 1-39.

10. Draw lines (wires) between the logic gates and the input and output symbols. From the **OPTIONS** menu select **Line Styles** and then the **Solid Line** (option at the top of the window). Place the selection pointer on the line at the end of the "Quarter1" input symbol. The selection pointer turns into a "+" shape drawing tool. While pressing the mouse button, drag the mouse to draw the line to connect the "Quarter1" input symbol to the input of the "AND" gate.

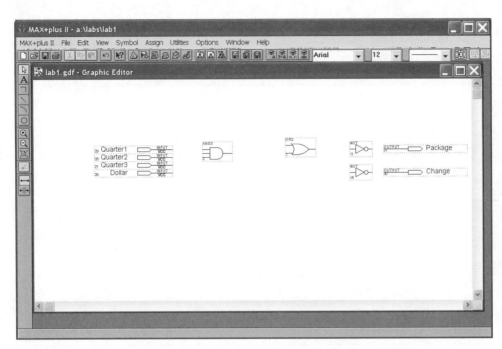

Figure 1-39 Altera "Graphic Editor" with Logic Gates, Input/Output Symbols, and Names

Repeat this process to draw lines connecting all the other input and output symbols. The Graphic Editor window should resemble Figure 1-40.

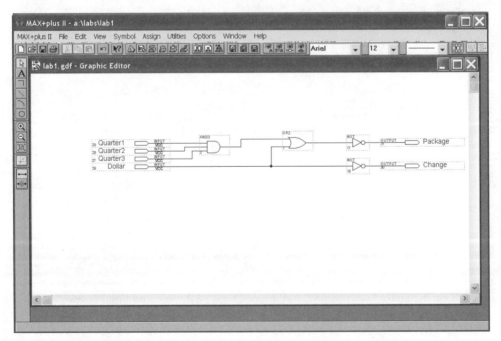

Figure 1-40 Altera "Graphic Editor" with Logic Gates, Input/Output Symbols, and Names and Lines

Here are some helpful tips that you can use to draw wires:

TIP #1: Green dotted box surrounds each symbol. This is a symbol boundary box. You must never run wires through (inside) the boundary box. You must always terminate connections at the stub end of a symbol's boundary box. Wires that run on top of a boundary box line, on the input or output stub, will automatically be connected to the symbol. You must provide a space between the wire and the symbol boundary box line when you run a wire past a symbol.

TIP #2: *To anchor a bend (or elbow) in the wire you must release the mouse button and reclick it. A nonanchored elbow will move around as you move the mouse.*

At this point, the diagram has been drawn and you are about to assign an Altera IC to the design. You must choose either the **MAX EPM7128S** or the **FLEX EPFX10K20** (or **FLEX EPF10K70** for the UP-2 board). This lab book refers to the **MAX EPM7128S** as the **MAX IC** and to the **FLEX EPFX10K20** (and **FLEX EPFX10K70**) as the **FLEX IC**. The instructor will tell you which IC to choose. There are two different sections for the next steps of the lab procedure. One section will teach you how to program the MAX IC, and the other section will teach you how to program the FLEX IC. To ensure you don't get mixed up, you will find **MAX** and **FLEX** subscript references on each step of the procedure.

Steps 11 through 18: Programming the MAX IC

If your instructor has decided to use the FLEX IC, then you should skip this section and proceed to the section titled "Steps 11 through 18: Programming the FLEX IC."

11_{MAX}. Select the **MAX IC**. The UP Board has two ICs. Your design will be programmed into the MAX IC. From the **ASSIGN** menu select **DEVICE**. Uncheck **Show only fastest speed grades** and then select **Device Family MAX7000S** and **Devices EPM7128SLC84-7**. Click the "OK" button. Visually, nothing changes on the worksheet. However, the MAX IC is now assigned to the design.

12_{MAX}. To ensure that your drawing is correct, you can use MAX+PLUS II to check your worksheet for basic errors. From the "**File**" menu select **Project Save & Check**. If the "Project Save & Check" menu item is grayed out (not selectable), you will need to select **Set Project to Current File** and try again to select the **Project Save & Check** menu item. The file is saved and the MAX+PLUS II "Compiler" window opens; the compiler checks the file for errors and displays a message window indicating the number of errors and warnings. The "Compiler" window (without the error message window) should resemble Figure 1-41.

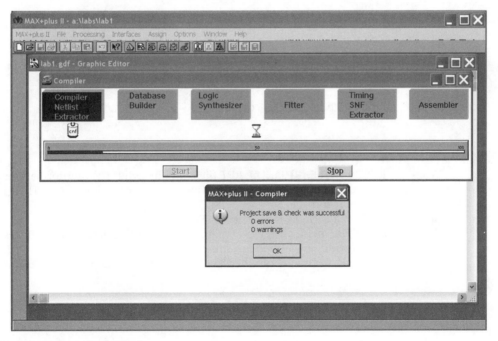

Figure 1-41 Altera "Compiler" Window

If the compiler issues any error messages, you should double-click on the first error message in the list. The MAX+plus II software will revert to the Graphic Editor and show the location of the error in *red*. Study the diagram and correct the problem. Repeat the "Save & Check" procedure. You may need to correct other errors and you may need to get help from the instructor to resolve the problems. For assistance, you can also click on the first error message in the list and then click on the "Help on Message" button.

When "Save & Check" is error free, close the compiler window (click the X button in the top right of the window). You will return to the Graphics Editor window and you can proceed with the lab.

13$_{MAX}$. To generate the output files for simulation and programming and to fit the logic design into the MAX IC, you will need to compile the project. From the **File** menu select **Project Save & Compile**. The file is saved and the MAX+PLUS II "Compiler" window opens; the compiler checks the file for compiler errors and creates the simulation and programming files. The same compiler window opens again, but this time the red progress bar goes completely across the screen to 100%. The compiler may display warning messages. The warning messages can be ignored. The "compiler" window (with the warning message window) should resemble Figure 1-42.

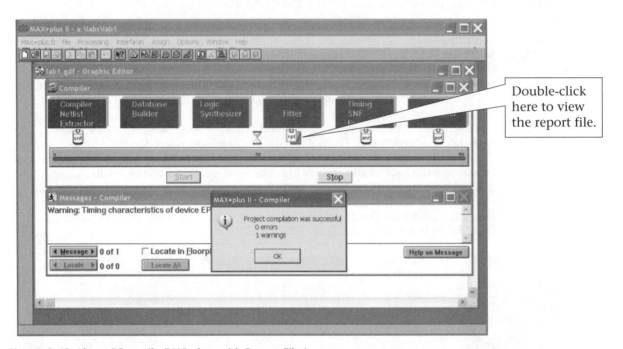

Figure 1-42 Altera "Compiler" Window with Report File Icon

View the **REPORT FILE (RPT)**. The **RPT** file contains two types of information: project device compilation messages and input/output connection information. There are two ways to view the report file:

a. Double-click the **RPT** file icon. Refer to Figure 1-42.
b. Click on the **File** menu and select **Open**. Open the file **Lab1.rpt**. In both cases, the "Text Editor" window will open to show the "RPT" file. Scroll down the RPT file screen until you see the IC outline. The "Text Editor" window should resemble Figure 1-43.

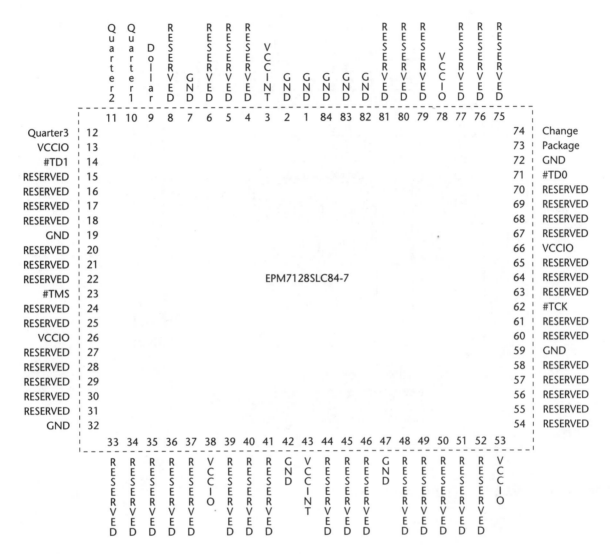

Figure 1-43 Altera "Report File" IC Outline

The square outline represents the MAX IC. The number 1 is at the top and center of IC. The numbers increase from 1 to 2 to 3 in a counterclockwise direction. The largest number, 84, completes the full circle around the IC and is beside number 1. These numbers represent connection points on the IC. The points are commonly referred to as pins, and the numbers are called pin numbers. The pins transfer binary data in and out of the MAX IC. The pins of the MAX IC are connected to a set of headers (P1, P2, P3, and P4) surrounding the IC. A header is a socket that will allow you to insert wires to make connections to the switches and LEDs. The diagram shown in Figure 1-44 can be used to easily locate the pin numbers on the MAX IC header sockets. Each pin number is shown in a box along with a numbered counting guide (1... to ... 6). The counting guide shows you the number of pins you must count off from each end of the header socket in order to make a wire connection.

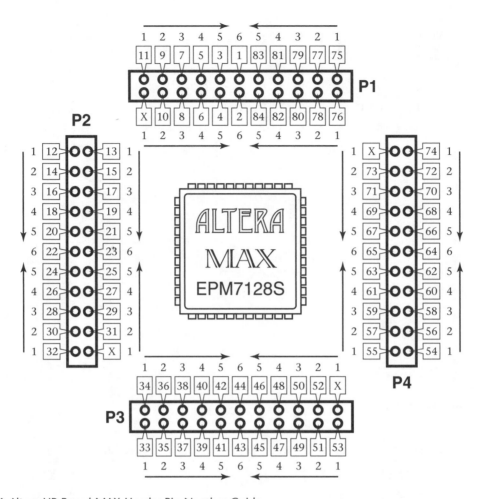

Figure 1-44 Altera UP Board MAX Header Pin Number Guide

14_{MAX}. You will connect four switches and two LEDs to the vending machine system. Begin with the switch connection. According to the report file input, **Quarter1** is pin number 10. According to the header pin number diagram, pin "10" is located on the inside edge of "P1" two pins in from the left edge of the header. Insert one end of a wire into pin "10" of the header socket and the other end into the header for a switch. Repeat this procedure to connect switches for inputs **Quarter2**, **Quarter3**, and **Dollar**. According to the report file input **Quarter2** is pin number **11**, **Quarter3** is pin number **12**, and **Dollar** is pin number **9**. Figure 1-45 shows the wire connections for the switches.

Connect an LED to the output "Package." According to the report file output, "Package" is pin number "73." According to the header pin number diagram, pin "73" is located on the inside edge of "P4" two pins from the top of the header. Insert one end of a wire into pin "73" of the header socket and the other end into the header for an LED. Repeat this procedure to connect an LED to output "Change." According to the report file, "Change" is pin number "74." Figure 1-45 shows the wire connections for the LEDs.

NOTE: *There is a possibility that your compiler generated different pin numbers than the ones shown in the diagram. That is not a problem. Substitute your pin numbers for the ones that are shown on Figure 1-43.*

15_{MAX}. Connect the BB cable (Byte Blaster) to the LPT1: printer port on your PC.

16_{MAX}. Connect the power pack to the UP board. The "power" LED will light on the UP board.

17_{MAX}. Program the MAX IC. This step will transfer the design into the MAX IC. From the **MAX+PLUS II** menu select **Programmer**. The window shown in Figure 1-46 will open.

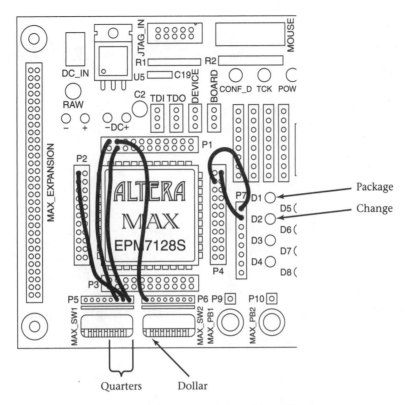

Figure 1-45 Connecting Switches and LEDs to the Altera UP Board Max Header

Figure 1-46 Altera "Programmer" Window

NOTE: *The file parameters for the last person to use the Altera Programmer are always saved. This may result in a "Can't open file" error message. This error message can be ignored.*

From the menu, that runs across the very top of the screen, click on **JTAG**. Make sure there is a check mark beside the "Multi-Device JTAG Chain" menu item. If the check mark is not there then select **Multi-Device JTAG Chain** to automatically place the check mark on this menu item. This action should open the "Multi-Device JTAG Chain" window shown in Figure 1-47. If not, then from the menu at the top of the screen, click on **JTAG**. Choose **Multi-Device JTAG Chain Setup**. The window shown in Figure 1-47 will open.

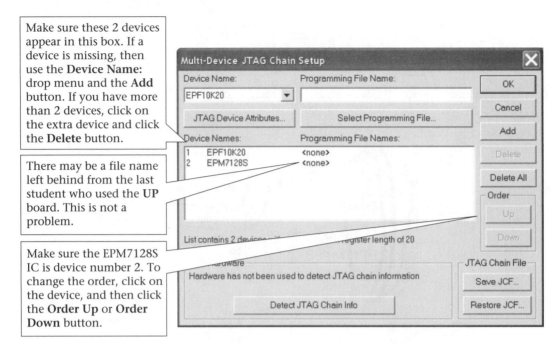

Make sure these 2 devices appear in this box. If a device is missing, then use the **Device Name:** drop menu and the **Add** button. If you have more than 2 devices, click on the extra device and click the **Delete** button.

There may be a file name left behind from the last student who used the **UP** board. This is not a problem.

Make sure the EPM7128S IC is device number 2. To change the order, click on the device, and then click the **Order Up** or **Order Down** button.

Figure 1-47 Altera "Multi-Device JTAG Chain Setup" Window

Follow this procedure:

a. Click on **device 2 EPM7128S**.

b. Click the **Select Programming File** button. A window will open that will allow you to select **lab1.pof**. Click the "OK" button.

c. When you return to the "Multi-Device JTAG Chain Setup" window, click the **Change** button and then click the "OK" button.

d. When you return to the "Programmer" window, click the **Program** button and the design will be transferred to the IC. Wait until the red progress bar goes completely across the screen to 100% before proceeding.

18_{MAX}. At this point you have drawn the diagram, programmed the IC, and connected the switches and LEDs. The system is ready for testing.

Test A. Flip all switches to the 0 position. This represents the condition where *no money* is inserted in the machine. Both LEDs should be *off*.

Test B. Flip only the three-quarter inputs to the 1 position. This test should result in the "Package" LED turning *on* and the "Change" LED *off*.

Test C. Flip only the **Dollar** input to the 1 position. This test should result in both the "Package" and "Change" LED turning *on*.

Steps 11 through 18: Programming the FLEX IC

If your instructor has decided to use the MAX IC, then you should skip this section. You have already completed steps 11 through 18 of the lab for the MAX IC and you do not need to complete these steps again for the FLEX IC.

11_{FLEX}. Select the **FLEX IC**. The UP Board has two ICs. Your design will be programmed into the FLEX IC. From the **ASSIGN** menu select **DEVICE**. Uncheck **Show only fastest speed grades**, and then select **Device Family FLEX10K** and **Devices EPF10K20RC240-4** (or **EPF10K70RC240-4** for the UP-2 board). Click the "OK" button. Visually, nothing changes on the worksheet. However, the FLEX IC is now assigned to the design.

12_{FLEX}. To ensure that your drawing is correct you can use MAX+PLUS II to check your worksheet for basic errors. From the **File** menu select **Project Save & Check**. If the "Project Save & Check" menu item is grayed out (not selectable), you will need to select **Set Project to Current File** and try again to select the **Project Save & Check** menu item. The file is saved and the MAX+PLUS II "Compiler" window opens; the compiler checks the file for errors and displays a message window indicating the number of errors and warnings. The compiler window (without the error message window) should resemble Figure 1-48.

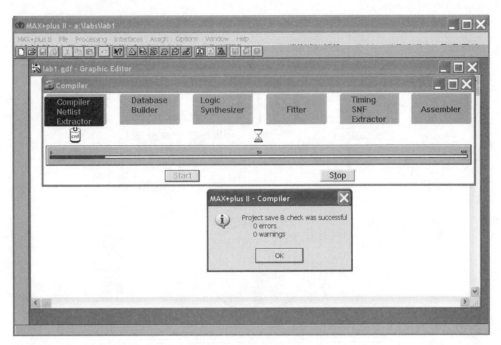

Figure 1-48 Altera "Compiler" Window

If the compiler issues any error messages, you should double-click on the first error message in the list. The MAX+PLUS II software will revert to the Graphic Editor and show the location of the error in red. Study the diagram and correct the problem. Repeat the "Save & Check" procedure. You may need to correct other errors, and you may need to get help from the instructor to resolve the problems. For assistance, you can also click on the first error message in the list and then click on the "Help on Message" button.

When Save & Check is error free, close the Compiler window (click the X button in the top right of the window). You will return to the Graphics Editor window and you can proceed with the lab.

13_{FLEX}. You need to connect four switches to the vending machine system. The FLEX IC has a set of eight onboard switches. Refer to Figure 1-49.

Printed circuit board wire connections (PCB traces) connect the switches to the FLEX IC. You do not need to connect external wires from the switches to the IC. The FLEX IC has 240 tiny connection points across 4 sides of the IC (60 per side). These shiny silver metal legs connect the IC to the PCB and transfer binary data in and out of the Altera IC. The legs are called **pins**. The *pins* are identified with pin numbers. Pins 1 through 60 are located on the left side of the IC. Eight of these pins are connected to switches. Refer to Figure 1-50.

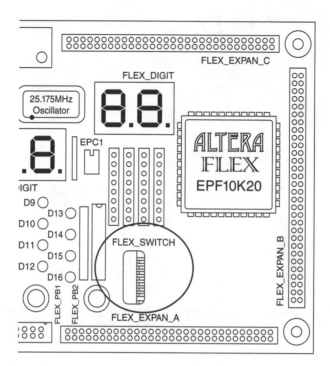

Figure 1-49 Altera UP Board FLEX Switches

FLEX Switch Pin Assignments

	EPF10K Pin #	
FLEX_SWITCH-1	41	Switch closest to the bottom of the UP board
FLEX_SWITCH-2	40	
FLEX_SWITCH-3	39	
FLEX_SWITCH-4	38	
FLEX_SWITCH-5	36	
FLEX_SWITCH-6	35	
FLEX_SWITCH-7	34	
FLEX_SWITCH-8	33	Switch closest to the top of the UP board

Figure 1-50 Altera UP Board FLEX Switch Pin Assignments

To connect the switches, you must assign a pin number to each input symbol. Place the selection pointer over the input symbol "Quarter1." Right click and select **Assign** then **Pin/Location/Chip**.... Refer to Figure 1-51. The "Pin/Location/Chip" window will appear. Refer to Figure 1-52. Type or select the number **Pin: 41** to assign **FLEX_SWITCH-1**. Then click the "Add" button. Refer to Figure 1-53. Click the "OK" button. Your design will now be able to connect the "Quarter1" input to FLEX_SWITCH-1.

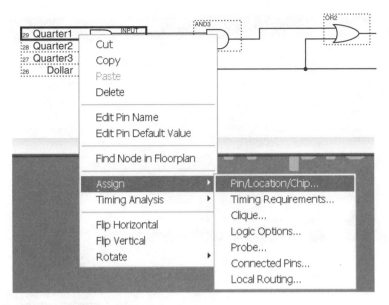

Figure 1-51 Altera "Assign Pin" Menu

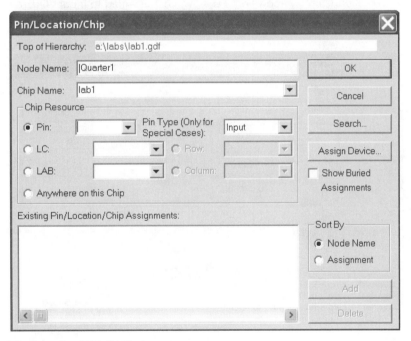

Figure 1-52 Altera "Pin/Location/Chip" Window

Place the selection pointer over each of the other input symbols "Quarter2," "Quarter3," and "Dollar" and repeat this procedure to assign pin numbers (40, 39, and 38) that will connect the remaining input symbols to switches 2, 3, and 4. Instead of pressing the "Add" button, click the "OK" button after each pin number entry. This will add the pin number, close the window, and allow you to quickly right click over the next input symbol. When you are done, the "Pin/Location/Chip" window should resemble Figure 1-54.

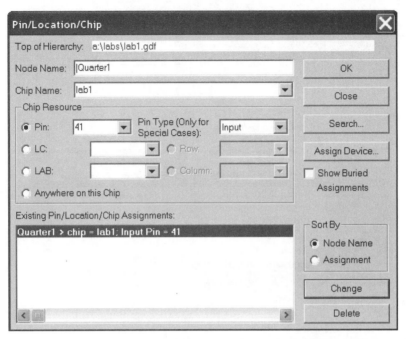

Figure 1-53 Altera "Pin/Location/Chip" window with FLEX Switch 1 assigned to input "Quarter1".

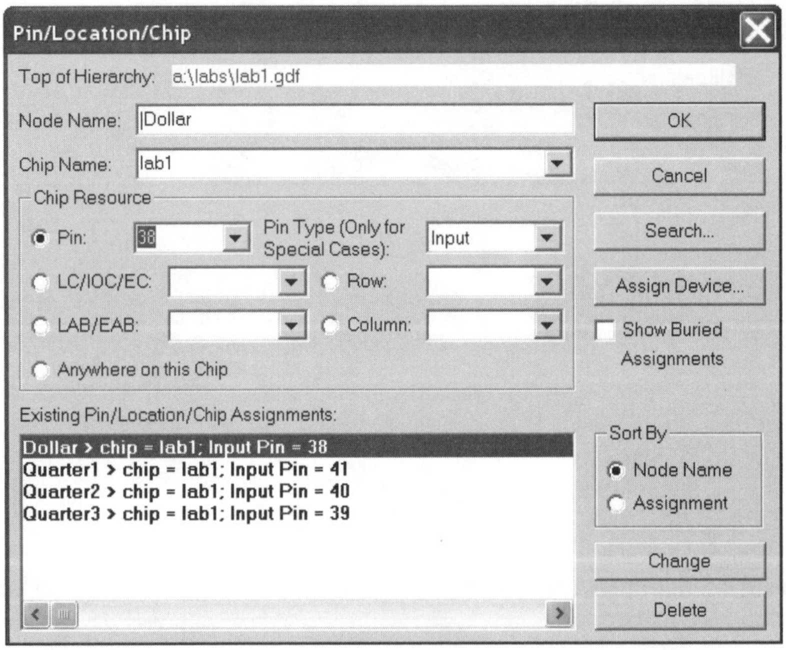

Figure 1-54 Altera "Pin/Location/Chip" window with FLEX switches assigned to all inputs

14$_{FLEX}$. At this point the pin numbers to connect the switches have been assigned. You need to assign pin numbers to connect the LEDs. The FLEX IC is connected to a pair of digital displays. Refer to Figure 1-55. Each digital display has eight active LOW LEDs. There are four vertical LEDs, three horizontal LEDs, and a decimal point. Refer to Figure 1-56. The horizontal LEDs can be turned off and the vertical LEDs can be used to create an 8-bit display. Refer to Figure 1-57. The LEDs are identified with labels D1, D2, D3, D4, D5, D6, D7, and D8. Refer to Figure 1-58.

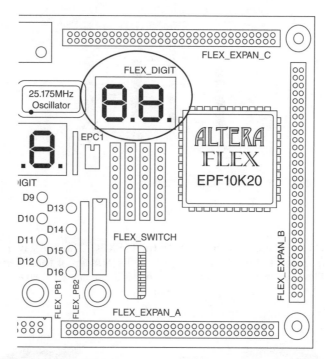

Figure 1-55 Altera UP Board FLEX Digital LED Displays

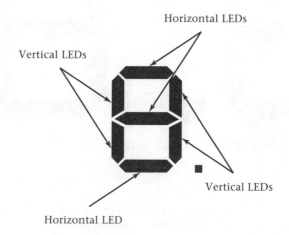

Figure 1-56 Altera UP Board FLEX LED Display Layout

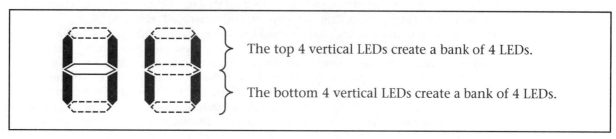

Figure 1-57 Altera UP Board FLEX LED Display as an 8-Bit Display

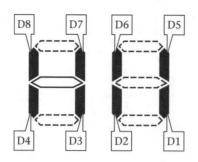

Figure 1-58 Altera UP board FLEX 8-bit LED Display Labels

PCB traces connect the LEDs to the FLEX IC. The FLEX IC has 240 pins. Pins 1 through 60 are located on the left side of the IC. Sixteen of these pins are connected to the LEDs. Figure 1-59 shows the pin numbers inside the boxes.

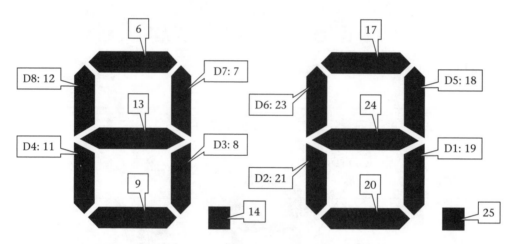

Figure 1-59 Altera UP Board FLEX 8-Bit LED Display Pin Numbers

LEDs D1 and D2 will be connected to outputs "Package" and "Change." To connect the LEDs you must assign a pin number to each output symbol. Place the selection pointer over the output symbol "**Package**," right click, and select **Assign** then **Pin/Location/Chip…** . Type or select the number **Pin: 19 (LED D1)**, then click the "Add" button. Refer to Figure 1-60.

Click the "OK" button. Place the selection pointer over the output symbol "Change " and repeat this procedure to assign the output symbol to **LED D2 (pin 21)**. When you are done, the "Pin/Location/Chip" window should resemble Figure 1-61.

You need to turn off all of the other unused LEDs in order to block the MAX+PLUS II software from randomly routing logic signals through these pins and turning them on. Having unused LEDs inadvertently turn on is a visual distraction. Active low LEDs can be turned off with a connection to logic 1. To turn off the 14 unused LEDs you must place 14 output symbols on the worksheet. Select an area to the right of the current drawing to place the output symbols. Connect the output symbols to a logic-1 symbol (the symbol name is "**Vcc**" in the "prim" library). Name each output symbol as shown in Figure 1-62.

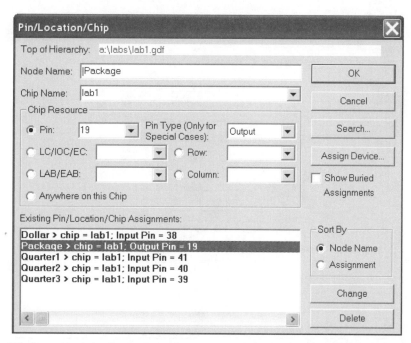

Figure 1-60 Altera "Pin/Location/Chip" Window with FLEX LED D1 assigned to the package output

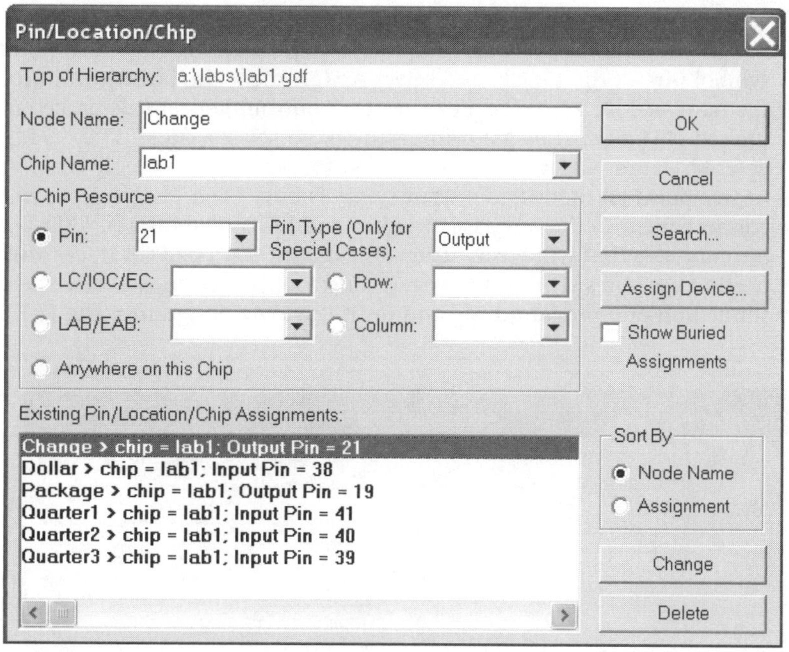

Figure 1-61 Altera "Pin/Location/Chip" Window with FLEX LEDs assigned to all outputs

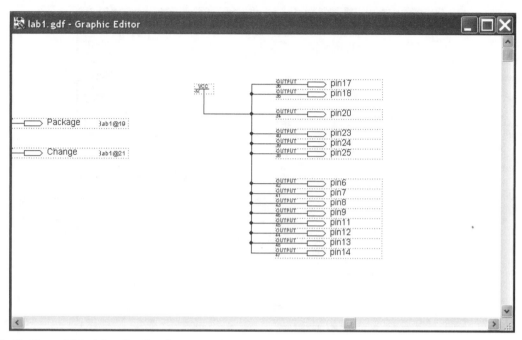

Figure 1-62 Altera "Graphic Editor" with Unused FLEX LEDs connected to 5 volts

Assign pin numbers to each "output symbol." Place the selection pointer over the input symbol **pin17**, right click, and select **Assign** then **Pin/Location/Chip...**. Type or select the number **Pin: 17**, then click the "OK" button. Repeat this procedure for the other unused LED pins. They are numbered 18, 20, 23, 24, 25, 6, 7, 8, 9, 11, 12, 13, and 14. Refer to Figure 1-63.

15_{FLEX}. At this point all input and output symbols have been assigned pin numbers that will connect them to the onboard switches and LEDs. The unused LEDs have been assigned pin numbers that will ensure they do *not* light up. The PCB traces will eliminate the need to use wires. The compile step is next. It is used to generate the output files for simulation and programming and to fit the logic design into the **FLEX IC**. Some PCs

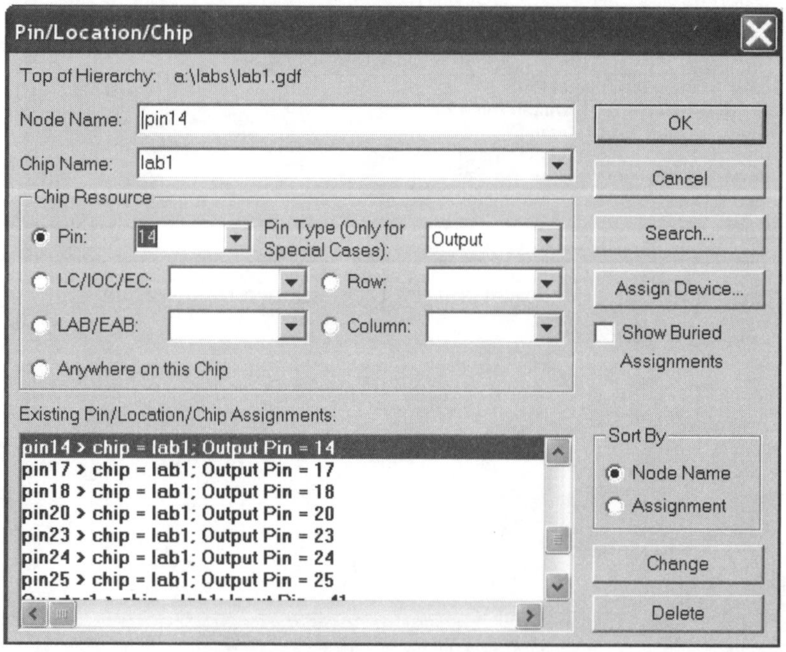

Figure 1-63 Altera "Pin/Location/Chip" window with pin numbers assigned to all unused FLEX LEDs

need the "Quartus Fitter" to be turned off to be able to compile without generating a fatal error. Other PCs will work with the "Quartus Fitter" *on*. The "Quartus Fitter" is *on* by default. To turn *off* the "Quartus Fitter" follow these steps:

From the "**MAX+PLUS II**" menu select **Compiler**. The "compiler" window will open and allow you to select the **Processing** menu. From the **Processing** menu select **Fitter Settings...** . This will open the "Fitter Settings" window. From the **Fitter Settings** window *uncheck* "**Use Quartus Fitter for FLEX10K and ACEX1k Devices**" and click "OK."

The "Quartus fitter" settings will be stored with the project files. You need only turn it off the very first time you compile. You are now ready to compile the project. From the **FILE** menu select **Project Save & Compile**. The file is saved and the MAX+PLUS II "Compiler" window opens; the compiler checks the file for compiler errors and creates the simulation and programming files. Wait for the red progress bar to go completely across the screen to 100%. The compiler may display warning messages. These warning messages can be ignored. The "compiler" window (with the warning message window) should resemble Figure 1-64.

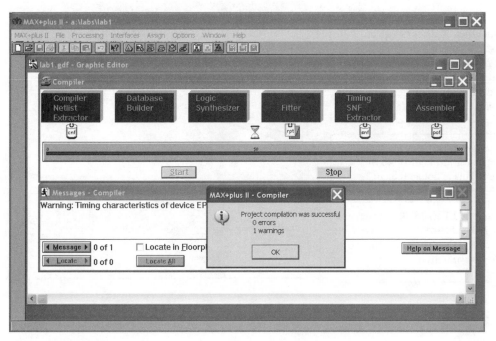

Figure 1-64 Altera "Compiler" Window

16_{FLEX}. You need to power the UP board in order to program the FLEX IC. Connect the BB cable (Byte Blaster) to the LPT1: printer port on your PC. Connect the power pack to the UP board. The "power" LED will light on the UP board.

17_{FLEX}. This step will transfer the design into the FLEX IC (program the FLEX IC). From the **MAX+PLUS II** menu select **Programmer**. The window shown in Figure 1-65 will open.

NOTE: *The file parameters for the last person to use the Altera Programmer are always saved. This may result in a "Can't open file" error message. This error message can be ignored.*

From the menu that runs across the very top of the screen, click on **JTAG**. Make sure there is a check mark beside the "Multi-Device JTAG Chain" menu item. If the check mark is not there, then select **Multi-Device JTAG Chain** to automatically place the check mark on this menu item. This action should open the "Multi-Device JTAG Chain" window shown in Figure 1-66. If not, then from the menu at the top of the screen, click on **JTAG**. Choose **Multi-Device JTAG Chain Setup**. The window shown in Figure 1-66 will open.

Figure 1-65 Altera "Programmer" Window

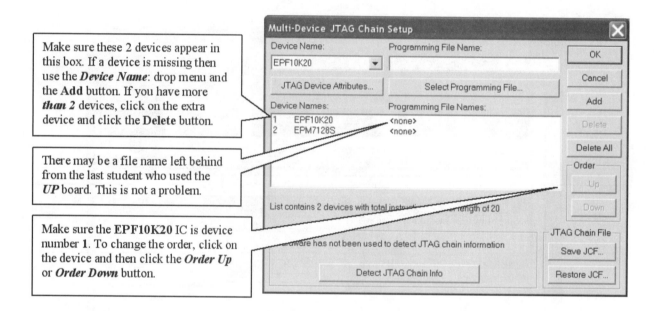

Make sure these 2 devices appear in this box. If a device is missing then use the **Device Name**: drop menu and the **Add** button. If you have more **than 2** devices, click on the extra device and click the **Delete** button.

There may be a file name left behind from the last student who used the **UP** board. This is not a problem.

Make sure the **EPF10K20** IC is device number **1**. To change the order, click on the device and then click the **Order Up** or **Order Down** button.

Figure 1-66 Altera "Multi-Device JTAG Chain Setup" Window

Follow this procedure:

a. Click on **device 1 "EPF10K20"** (or **EPF10K70** for the UP-2 board).

b. Click the **Select Programming File** button. A window will open that will allow you to select **lab1.sof**. Click the "OK" button.

c. When you return to the "Multi-Device JTAG Chain Setup" dialog window, click the **Change** button and then click the "OK" button.

d. When you return to the "Programmer" dialog window, click the **Configure** button and the design will be transferred to the IC. Wait until the red progress bar goes completely across the screen to 100% before proceeding.

18_{FLEX}. At this point you have drawn the diagram, programmed the IC, and connected the switches and LEDs. The system is ready for testing.

Test A. Flip all switches to the 0 position. This represents the condition where no money is inserted in the machine. Both LEDs should be *off*.

Test B. Flip only the three-quarter inputs to the 1 position. This test should result in the "Package" LED turning *on* and the "Change" LED turning *off*.

Test C. Flip only the "Dollar" input to the 1 position. This test should result in both the "Package" and "Change" LED turning *on*.

LAB EXERCISES AND QUESTIONS

This section contains lab exercises that can be performed on the UP board and questions that can be answered at home. Ask your instructor which exercises to perform and which questions to answer.

Lab Exercises

Exercise 1: Generate a four-input and two-output truth table to record the system response.

Generate a four-input and two-output truth table to record the response of the vending machine system.

Exercise 2: The Altera Simulator: (This exercise can be done at home.)

The information presented in Appendix F will teach you how to use the **Altera Max+Plus II simulator**. A simulator is a program that allows you to test a design without having to program an IC. A simulation can help find design errors. The steps required to use the simulator will be applied to the vending machine system. Proceed to Appendix F and complete the simulation work.

Exercise 3: Improve the vending machine system.

a. The vending machine system has flaws in the design. Here is a scenario that would frustrate a user of the vending machine: A user searches his pocket, finds a quarter, and inserts it into the machine. Then he realizes that he does not have two more quarters but he does have a dollar! If he inserts the dollar, what would be the result? Test this condition on the vending machine system before answering the question.

b. Here is another scenario that would frustrate a user of the vending machine: A user searches his pocket, finds two quarters, and inserts them into the machine. Then he realizes that he does not have one more quarter but he does have a dollar! If he inserts the dollar, what would be the result? Test this condition on the vending machine system before answering the question.

c. To correct the problems described in a and b, you can add additional change outputs ("**Change2**," "**Change3**"). Improve the vending machine by adding extra logic gates and output LEDs to the design. Before testing the new system, some important points should be made regarding the UP board. Whenever you make a design change, you must recheck for basic errors with the "Project Save & Check" procedure. You must regenerate a new report and programming files with the "Project Save & Compile" procedure. For a MAX IC design, you must also review the report file to verify the pin numbers. The compiler can reassign MAX pin numbers each time you recompile. You may need to move MAX switch and MAX LED wires. Finally, the MAX IC and FLEX IC must be reprogrammed. Test the system and record the results in a truth table with four inputs and four outputs.

Lab Questions

1. Use the binary PWC to convert the following binary numbers to decimal:
   a. 101
   b. 1010
   c. 101011
   d. 110101
   e. 1011011
2. Use the binary PWC to convert the following decimal numbers to binary:
   a. 12
   b. 31
   c. 22
   d. 70
   e. 100
3. Draw the logic symbol for a four-input OR gate. Draw the truth table for a four-input OR gate. Write the Boolean equation for a four-input OR gate.
4. Draw the logic gate diagram for $X = A \cdot B + C \cdot D$

   HINT: *Look for P-terms and consider them to be a block. This will tell you what the output gate is. From there draw the gates for the P-terms.*

5. Design a digital system that can be used in a council meeting to determine if a majority of three council members agree on passing a resolution. Each council member sits at a table and has a switch that they can flip to the "Yes" position or to the "No" position. At the front of the council chamber, there are two lights (LEDs). One is labeled "Pass the resolution" and the other is labeled "Reject the resolution." If at least two of the three members agree (switch in the "Yes" position), then light the "Pass" LED, or else, light the "Reject" LED. Draw the logic gate diagram.

HINT *Declare your I/O. Try to write the Boolean equation for the system from the description.*

6. Each of the following statements describes a logic gate. From the statement, draw the logic gate symbol and name the symbol.
   a. The only time the output will be low is if both inputs are low.
   b. The output will be low if one or both inputs are high.
   c. The only time the output will be low is if both inputs are high.
   d. The output will be high if one or both inputs are high.
   e. The output will be high if the two inputs have the opposite logic level.

7. Generate a truth table for each symbol shown in Figure 1-67.

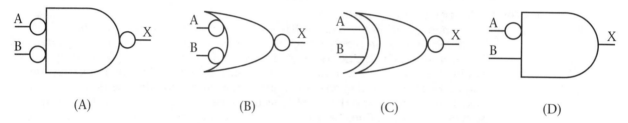

(A)   (B)   (C)   (D)

Figure 1-67 Symbols for Question 7

CONCLUSIONS

- A PWC can be used to convert decimal numbers to binary.
- Switches and LEDs on the Altera UP board represent input and output signals of a digital system.
- The three gates that make up the fundamental building blocks of all digital systems are called the NOT gate, the AND gate, and the OR gate.
- The UP board can be used to test the operation of a vending machine system.
- Design changes using the UP board are easily accommodated as long as you remember to recheck and recompile the project and reprogram the IC.
- An entire design requiring several older generation ICs can easily be placed inside a single Altera IC.
- A simulator is a program that allows you to test the operation of a digital system without having to program a CPLD. It can be used to find design errors.

Lab 2

Older Generation Logic Gate Systems Versus CPLD Systems

Equipment
Altera UP board
Blank floppy disk
Book CD-ROM.
Spool of 24 AWG wire
Wire cutters

Objectives
Upon completion of this lab, you should be able to:
- Compare the design cycle of older generation IC systems to that of CPLD systems.
- Convert an older generation IC system to a CPLD system.
- Use VHDL to design a digital system.

INTRODUCTORY INFORMATION

In Lab 1 you learned how to connect logic gate symbols to create a vending machine system. In this lab you will learn techniques that can be used to analyze and convert older generation digital systems to Altera CPLD systems. You will also learn how to use VHDL to design digital systems. VHDL is an industry standard hardware programming language that can simplify the process of creating a digital system.

VHDL Fundamentals

VHDL is an acronym for the **VHSIC Hardware Description Language**. **VHSIC** is an acronym for **Very High Speed Integrated Circuit**. VHDL is a text-based method of describing a digital system for Altera CPLD ICs. VHDL is also a widely accepted industry standard. Here is the vending machine from Lab 1 defined in VHDL. Refer to Figure 2-1.

```
vending.vhd - Text Editor                                    _ □ X
LIBRARY ieee;
USE ieee.STD_LOGIC_1164.ALL;


ENTITY vending IS
    PORT(
        Quarter1, Quarter2, Quarter3, Dollar    : IN    STD_LOGIC;
        Pack, Change                            : OUT   STD_LOGIC);
END vending;


ARCHITECTURE a OF vending IS
Begin
    Pack <= (Quarter1 and Quarter2 and Quarter3) or Dollar;
    Change <= Dollar;
End a;

Line  12   Col  9   INS ◄
```

Figure 2-1 VHDL Vending Machine System

Which method should you use to design a digital system? When students are introduced to digital systems, using logic gate symbols with the Graphic Editor is the best method. Pictures are easier to understand than words. Students can concentrate on digital concepts without struggling with VHDL syntax (punctuation or grammar rules). Once concepts are learned, VHDL becomes more advantageous. Think of a complex design that is made up of over 100 logic gates. The diagram in the Graphic Editor becomes large, complicated to draw, and difficult to analyze visually. There is no advantage to seeing a picture of the entire system when it becomes large and cluttered. The same design defined in VHDL can be organized into modules or blocks of text. Each block of text is an entity that can be easily analyzed and changed. VHDL modules can also be made to appear as a symbol in the Graphic Editor. A digital system can be made up as a mix of standard symbols and VHDL modules. Organizing complex designs this way is easier using VHDL.

This book takes advantage of both design entry methods. Graphic Editor design entry is better suited to learning digital system design. VHDL makes complex system designs easier to manage.

There are two structures in a VHDL design: an **entity** declaration and an **architecture** body. The entity is the box with input and output definitions and the architecture is what's inside the box to make it work. The entity is like the framework or chassis of a car, and the architecture is like the engine of a car. Refer to Figure 2-2.

VHDL Selected Signal Assignment

A small Boolean equation is easy to enter as a VHDL statement. A complex Boolean equation is more difficult. VHDL has a structure that allows you to enter a digital design as a truth table. A truth table clearly shows a digital system's response and is a more intuitive method of creating a digital system. The name of the VHDL structure is **Selected Signal Assignment (VHDL SSA)**. Compare the truth table and the VHDL statements for the vending machine. Refer to Figure 2-3.

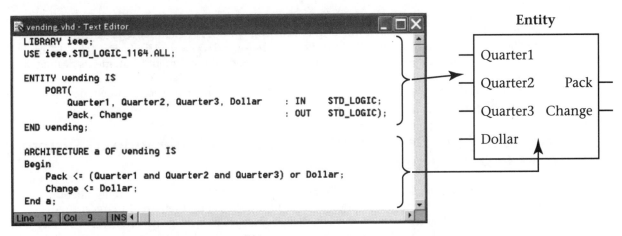

Figure 2-2 VHDL Entity Declaration and Architecture Body

Truth Table

L	Q3	Q2	Q1	P	C
0	0	0	0	0	0
0	0	0	1	0	0
0	0	1	0	0	0
0	0	1	1	0	0
0	1	0	0	0	0
0	1	0	1	0	0
0	1	1	0	0	0
0	1	1	1	1	0
1	0	0	0	1	1
1	0	0	1	1	1
1	0	1	0	1	1
1	0	1	1	1	1
1	1	0	0	1	1
1	1	0	1	1	1
1	1	1	0	1	1
1	1	1	1	1	1

```
LIBRARY ieee;
USE ieee.STD_LOGIC_1164.ALL;
ENTITY vending IS
        PORT(
                Q1, Q2, Q3, L          : IN    STD_LOGIC;
                P, C                    : OUT   STD_LOGIC);
END vending;

ARCHITECTURE a OF vending IS
        SIGNAL input: STD_LOGIC_VECTOR (3 DOWNTO 0);
        SIGNAL output: STD_LOGIC_VECTOR (1 DOWNTO 0);
BEGIN
-- Concurrent Signal Assignment
input   (3)   <= L;
input   (2)   <= Q3;
input   (1)   <= Q2;
input   (0)   <= Q1;

-- Selected Signal Assignment
WITH input SELECT
                output   <=
                                "00" WHEN "0000",
                                "00" WHEN "0001",
                                "00" WHEN "0010",
                                "00" WHEN "0011",

                                "00" WHEN "0100",
                                "00" WHEN "0101",
                                "00" WHEN "0110",
                                "10" WHEN "0111",

                                "11" WHEN "1000",
                                "11" WHEN "1001",
                                "11" WHEN "1010",
                                "11" WHEN "1011",

                                "11" WHEN "1100",
                                "11" WHEN "1101",
                                "11" WHEN "1110",
                                "11" WHEN "1111",

                                "00" WHEN others;

P       <=      output(1);
C       <=      output(0);
END a;
```

Input:

(3)	(2)	(1)	(0)
L	Q3	Q2	Q1

P	C	L	Q3	Q2	Q1

Output:

(1)	(0)
P	C

Q1, Q2, Q3 are quarters.
L is a dollar.
P is package.
C is change.

Figure 2-3 VHDL "Selected Signal Assignment" Structure for the Vending Machine System

The architecture uses a vector statement **SIGNAL input: STD_LOGIC_VECTOR (3 DOWNTO 0);** to define a storage array. The storage array is called **input** and (3 DOWNTO 0) implies it stores 4 bits of data. The statements input (3) <= L;, **input (2) <= Q3;, input (1) <= Q2;,** and **input (0) <= Q1;** assign the input variables to the storage array.

The architecture uses a vector statement **SIGNAL output: STD_LOGIC_VECTOR (1 DOWNTO 0);** to define a storage array. The storage array is called **output** and (1 DOWNTO 0) implies it stores 2 bits of data. The statements **P <= output(1);** and **C <= output(0);** assign the output array to the output variables P and C.

The "Selected Signal Assignment" statements **WITH input SELECT** will evaluate the input array (L, Q3, Q2, Q1) and assign the output array to *P* and *C*. For example, the statement **output <= "00" WHEN "0000"**, assigns *P = 0* and *C = 0* when *L = 0*, *Q3 = 0*, *Q2 = 0*, and *Q1 = 0*. This is the correct system response. The remaining "Selected Signal Assignment" statements evaluate the remaining truth table combinations to correctly assign values to *P* and *C*.

The structure of the "Select Signal Assignment" statements makes logic system design intuitively easy.

Comparing the Design Cycle of an Older Generation IC System to a CPLD IC System

Lab 1 described an older generation IC technology called **TTL**. TTL ICs are single-function ICs that are pieced together to create a design. Complex designs often require dozens (at times hundreds) of single-purpose ICs to build a digital system. The wiring was often very complex and messy. The newer Altera **MAX** and **FLEX** CPLD ICs can easily accommodate a complex digital system design that would require hundreds of older generation TTL ICs. To build and test the operation of a digital system using Altera CPLD ICs requires little wiring. All you need to connect are the input and the output devices to the Altera CPLD IC. Design changes or revisions are also very simple. Older generation IC technology would require a complete teardown and rebuild when changes were required. Altera CPLD ICs are easily reprogrammed.

To further illustrate the benefits of Altera CPLD IC technology, let's take a look at a typical design cycle for a digital system that is constructed using TTL ICs.

Step 1: The designer declared the inputs and outputs of the system.

Step 2: The designer generated a Boolean equation for the system.

The equation could be generated from the system truth table or directly from the system description.

Step 3: The designer used Boolean minimization techniques to minimize the size of the equation.

A smaller equation requires fewer gates, thus fewer TTL ICs to implement the design.

Step 4: The designer drew a logic gate diagram.

Step 5: The designer used a TTL IC data book to find the single-function logic gate ICs that were required to construct the system.

Step 6: A schematic diagram would be drawn showing the logic gate symbols and IC pin numbers.

Step 7: The designer purchased the ICs, and the schematic diagram was used to build a prototype.

TTL ICs are inexpensive; however, design changes usually result in a complete teardown and reconstruction of the system. You would need to repeat steps 2 through 7 to make design changes.

Step 8: The designer created and mass-produced a printed circuit board to mount all the ICs.

To illustrate steps 3, 5, and 6, Figure 2-4 shows a minimized Boolean equation, the TTL data book information, and the schematic diagram for a drill machine. To illustrate step 7, Figure 2-5 shows the wire connections necessary to prototype the drill machine.

$$X = D \bullet [\overline{A} \bullet \overline{B} + \overline{C} \bullet (\overline{A \bullet B})]$$

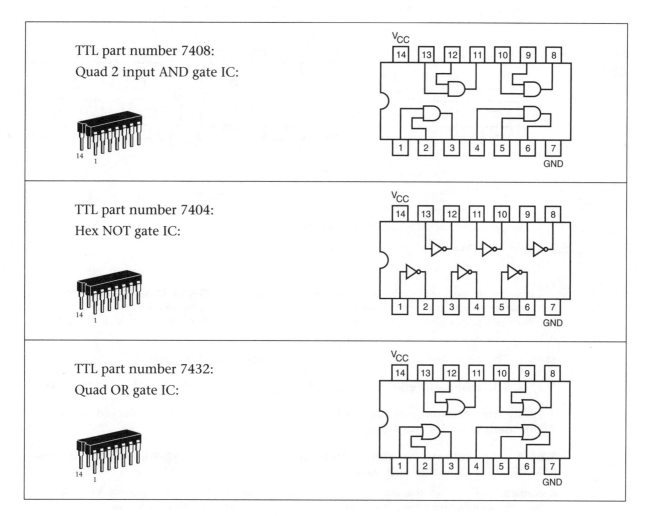

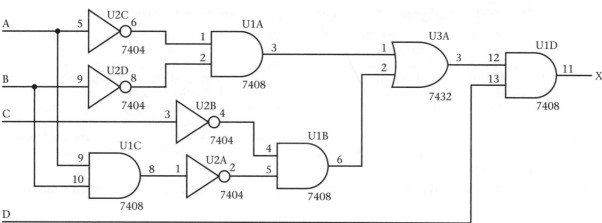

Figure 2-4 TTL IC Design Cycle: Drill Machine Equation and Schematic

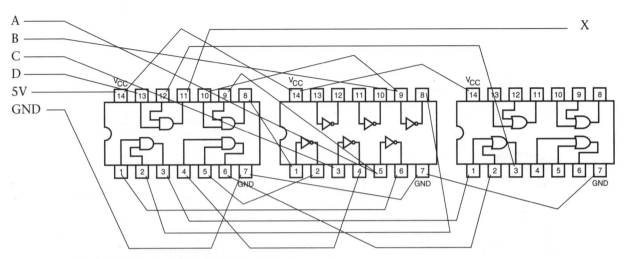

Figure 2-5 TTL IC Design Cycle: Drill Machine Wire Connections

Imagine the wiring complexity of a system more complicated than this simple drill machine. For a small increase in the prototyping cost, Altera CPLD ICs eliminates most of these problems.

Now let's take a look at design cycle of a digital system that is constructed with Altera CPLD ICs.

Step 1: The designer declares the inputs and outputs of the system.

Step 2: The designer generates a system truth table. No minimization is required.

Step 3: The designer uses the truth table to generate the VHDL Selected Signal Assignment code for the system.

Step 4: The designer can choose to simulate the operation of the system using the MAX+PLUS II simulator software.

A **simulator** is a program that allows you to test a design before it is programmed into an IC. A simulation can help find errors. Appendix F describes how to use the Altera "simulator."

Step 5: The designer tests the system by programming an Altera CPLD IC.

Altera CPLD ICs come in many different package configurations and the ideal IC solution may not be apparent in the early stages of system prototyping. A visit to the Altera Web site will allow you to locate the ideal IC for the design. Altera offers other types of system development kits (SDK). Think of an SDK as an industrial strength UP board. Altera offers different variations of SDKs for different variations of their programmable devices. The UP board is an effective prototyping system for educational institutions, but an SDK may be more beneficial when designing systems to be mass-produced. The prototyping step for TTL is less expensive because you don't need to buy a system development kit. However, Altera CPLD IC design changes are performed quickly by reprogramming the Altera CPLD IC. Design changes in TTL are very time consuming and the initial cost savings soon disappear.

Step 6: The designer creates and mass-produces a printed circuit board to mount the single IC design.

A visit to the Altera Web site will help you with this final step. The printed circuit board for a TTL design is usually larger and, thus, more expensive. This is another benefit to Altera CPLD ICs.

Converting an Old IC Technology System to a CPLD IC System

When working in the field you may encounter older generation IC technology systems, and you may need to convert the older system to an Altera CPLD IC system. The conversion process you use depends on the documentation that is available with the older system. Old systems are generally documented three different ways. Refer to Figure 2-6.

Truth Table

L	Q3	Q2	Q1	P	C
0	0	0	0	0	0
0	0	0	1	0	0
0	0	1	0	0	0
0	0	1	1	0	0
0	1	0	0	0	0
0	1	0	1	0	0
0	1	1	0	0	0
0	1	1	1	1	0
1	0	0	0	1	1
1	0	0	1	1	1
1	0	1	0	1	1
1	0	1	1	1	1
1	1	0	0	1	1
1	1	0	1	1	1
1	1	1	0	1	1
1	1	1	1	1	1

1—Truth table

$$P = Q1 \bullet Q2 \bullet Q3 + L$$
$$C = L$$

2—Boolean equation

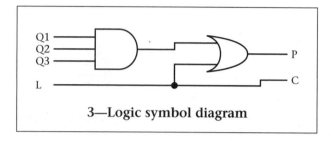

3—Logic symbol diagram

Figure 2-6 Older Generation IC System Documentation

Some older systems come with all three forms of documentation. Some older systems come with only one or two forms of documentation, whereas others may come with no documentation. If some form of documentation is available, then the procedures outlined in the next section will guide you through the conversion process. If no documentation is available, then you must design a functionally identical system from scratch. This is usually a little more time consuming. Another option is to remove the old ICs and trace the connections on the printed circuit board using an ohmmeter in order to re-create the system diagram. This, too, is time consuming.

You have seen how the VHDL selected signal assignment (VHDL SSA) structure can be used to easily create a digital system from a truth table. If the truth table for the older generation IC system is available, then the conversion process is simple. If the truth table is not available, then you will need to create the truth table from the existing system documentation. For this reason, you will learn conversion techniques that will allow you to generate the truth table for older generation system.

Old IC Technology System Defined by a Truth Table

An old IC technology system that includes truth table documentation is simple to convert to Altera CPLD IC technology. Use the truth table to generate the VHDL SSA code. Conversion techniques are not necessary.

Old IC Technology System Defined by a Boolean Equation

Assume that the old IC technology system includes only the Boolean equation documentation. You must know how to convert the Boolean equation to a truth table in order to use VHDL SSA. There are two techniques that can be used to generate a truth table from a Boolean equation: Technique 1: Equation substitution and Technique 2: Generate a truth table for sections of the equation.

Technique 1: Equation Substitution

Generate the truth table for the design. Substitute each entry into the equation and solve for X. Start with A,B,C = 0,0,0. See the example in Figure 2-7. Repeat the substitution procedure for each entry in the truth table.

$$X = \overline{\overline{A} \cdot B \cdot \overline{C}} + \overline{\overline{A} + \overline{C}}$$

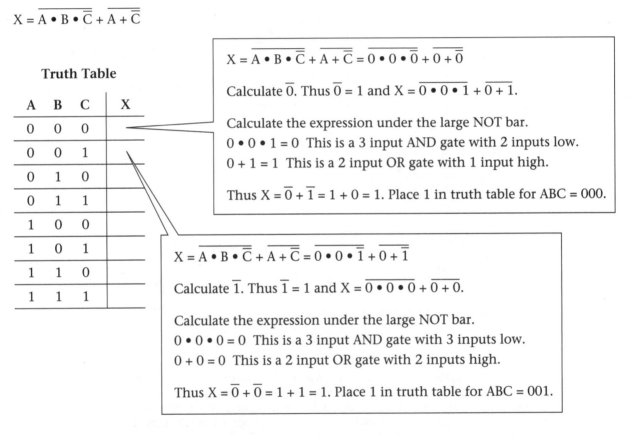

Truth Table

A	B	C	X
0	0	0	
0	0	1	
0	1	0	
0	1	1	
1	0	0	
1	0	1	
1	1	0	
1	1	1	

$X = \overline{\overline{A} \cdot B \cdot \overline{C}} + \overline{\overline{A} + \overline{C}} = \overline{\overline{0} \cdot 0 \cdot \overline{0}} + \overline{\overline{0} + \overline{0}}$

Calculate $\overline{0}$. Thus $\overline{0} = 1$ and $X = \overline{0 \cdot 0 \cdot 1} + \overline{0 + 1}$.

Calculate the expression under the large NOT bar.

$0 \cdot 0 \cdot 1 = 0$ This is a 3 input AND gate with 2 inputs low.
$0 + 1 = 1$ This is a 2 input OR gate with 1 input high.

Thus $X = \overline{0} + \overline{1} = 1 + 0 = 1$. Place 1 in truth table for ABC = 000.

$X = \overline{\overline{A} \cdot B \cdot \overline{C}} + \overline{\overline{A} + \overline{C}} = \overline{\overline{0} \cdot 0 \cdot \overline{1}} + \overline{\overline{0} + \overline{1}}$

Calculate $\overline{1}$. Thus $\overline{1} = 1$ and $X = \overline{0 \cdot 0 \cdot 0} + \overline{0 + 0}$.

Calculate the expression under the large NOT bar.

$0 \cdot 0 \cdot 0 = 0$ This is a 3 input AND gate with 3 inputs low.
$0 + 0 = 0$ This is a 2 input OR gate with 2 inputs high.

Thus $X = \overline{0} + \overline{0} = 1 + 1 = 1$. Place 1 in truth table for ABC = 001.

Figure 2-7 Generate a truth table for the equation: $X = \overline{\overline{A} \cdot B \cdot \overline{C}} + \overline{\overline{A} + \overline{C}}$.

Technique 2: Generate a Truth Table for Sections of the Equation

The following example demonstrates the steps that are required to use this conversion technique.
Example: $X = \overline{\overline{A} \cdot B \cdot \overline{C}} + \overline{\overline{A} + \overline{C}}$.

1. Generate a truth table with a wide area for the output response. Refer to Figure 2-8.
2. Generate a column for each product term (to review **P-term** see Lab 1). Start with the first P-term $\overline{A} \cdot B \cdot \overline{C}$. Ignore the large inversion bar $\overline{A} \cdot B \cdot \overline{C}$. Analyze the P-term and fill in the column. Place a 1 where A is high and B is high and C is low (because of the NOT operator on C). Fill in the remaining rows with 0s. Refer to Figure 2-9.
3. Generate a column for the second P-term $\overline{A} + \overline{C}$. Although the second term uses a + operator it is treated as a P-term because of the large inversion bar. Once again ignore the large inversion bar $\overline{A} + \overline{C}$. Analyze the P-term and fill in the column. Place a 1 where A is high or C is low. Fill in the remaining columns with 0s. Refer to Figure 2-10.

Truth Table

A	B	C		X
0	0	0		
0	0	1		
0	1	0		
0	1	1		
1	0	0		
1	0	1		
1	1	0		
1	1	1		

Figure 2-8 Generate a truth table for $X = \overline{A \cdot B \cdot \overline{C}} + \overline{A + \overline{C}}$. Draw a wide truth table.

Truth Table

A	B	C	$A \cdot B \cdot \overline{C}$		X
0	0	0	0		
0	0	1	0		
0	1	0	0		
0	1	1	0		
1	0	0	0		
1	0	1	0		
1	**1**	**0**	**1**		
1	1	1	0		

Figure 2-9 Generate a truth table for $X = \overline{A \cdot B \cdot \overline{C}} + \overline{A + \overline{C}}$. Generate a column for $A \cdot B \cdot \overline{C}$.

Truth Table

A	B	C	$A \cdot B \cdot \overline{C}$	$A + \overline{C}$		X
0	0	0	0	1		
0	0	1	0	0		
0	1	0	0	1		
0	1	1	0	0		
1	0	0	0	1		
1	0	1	0	1		
1	1	0	1	1		
1	1	1	0	1		

Figure 2-10 Generate a Truth Table for $X = \overline{A \cdot B \cdot \overline{C}} + \overline{A + \overline{C}}$. Generate a Column for $A + \overline{C}$.

Truth Table

A	B	C	$A \cdot B \cdot \overline{C}$	$A + \overline{C}$	$\overline{A \cdot B \cdot \overline{C}}$	$\overline{A + \overline{C}}$		X
0	0	0	0	1	**1**	0		
0	0	1	0	0	**1**	1		
0	1	0	0	1	**1**	0		
0	1	1	0	0	**1**	1		
1	0	0	0	1	**1**	0		
1	0	1	0	1	**1**	0		
1	1	0	1	1	**0**	0		
1	1	1	0	1	**1**	0		

Figure 2-11 Generate a truth table for $X = \overline{A \cdot B \cdot \overline{C}} + \overline{A + \overline{C}}$. Generate columns for $\overline{A \cdot B \cdot \overline{C}}$ and $\overline{A + \overline{C}}$.

4. Generate columns to include the inversion bars. Invert each bit from the original column. Refer to Figure 2-11.
5. The final step will generate column X. From the equation $X = \overline{A \cdot B \cdot \overline{C}} + \overline{A + \overline{C}}$, the two partial equations are separated by the OR operator. You must OR the two columns representing the partial equations. Refer to Figure 2-12.

Truth Table

A	B	C	$A \cdot B \cdot \overline{C}$	$A + \overline{C}$	$\overline{A \cdot B \cdot \overline{C}}$	$\overline{A + \overline{C}}$		X
0	0	0	0	1	**1**	0	1 + 0 =	1
0	0	1	0	0	**1**	1	1 + 1 =	1
0	1	0	0	1	**1**	0	1 + 0 =	1
0	1	1	0	0	**1**	1	1 + 1 =	1
1	0	0	0	1	**1**	0	1 + 0 =	1
1	0	1	0	1	**1**	0	1 + 0 =	1
1	1	0	1	1	**0**	0	0 + 0 =	0
1	1	1	0	1	**1**	0	1 + 0 =	1

Figure 2-12 Generate a truth table for $X = \overline{A \cdot B \cdot \overline{C}} + \overline{A + \overline{C}}$. OR the two columns.

Old IC Technology System Defined by a System Diagram

Assume that the old IC technology system includes only the system diagram documentation. You must know how to convert the system diagram to a truth table in order to use VHDL SSA. An old IC technology system that is defined by a system diagram can be converted to an Altera CPLD IC design two ways.

Option 1: If the system diagram is simple, then you may not need to convert the system diagram into a truth table. You can use the Altera Graphic Editor to interconnect logic gate symbols to draw the system diagram. This option is only recommended for simple diagrams that are simple to draw using the Graphic Editor.

Option 2: If you would want to use VHDL SSA to design the system, then you will have to convert the system diagram to a truth table. This option requires you to learn techniques to convert a system diagram to a truth table. The following example demonstrates the steps that are required to use such a conversion technique.

EXAMPLE: Convert a system diagram to a truth table. Refer to Figure 2-13.

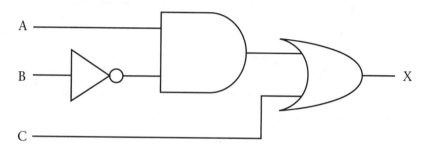

Figure 2-13 Convert a diagram to a truth table.

Generate a truth table and place each entry of the truth table onto the diagram and work out the response of the system. Start with ABC = 000. Apply it to the input of diagram. Refer to Figure 2-14. Write the logic gate response for each gate, working from the left side of the diagram to the right side. Record the value of X in the truth table. Refer to Figure 2-15.

Truth Table

A	B	C	X
0	0	0	
0	0	1	
0	1	0	
0	1	1	
1	0	0	
1	0	1	
1	1	0	
1	1	1	

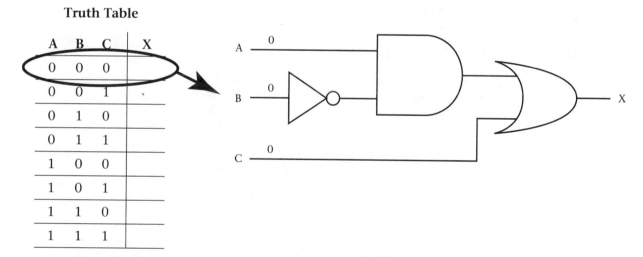

Figure 2-14 Convert a diagram to a truth table: Apply ABC = 000 to the input of diagram.

Truth Table

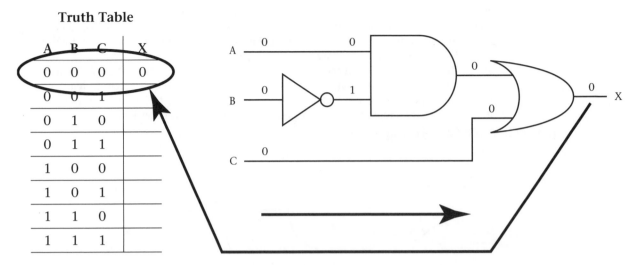

A	B	C	X
0	0	0	0
0	0	1	
0	1	0	
0	1	1	
1	0	0	
1	0	1	
1	1	0	
1	1	1	

Figure 2-15 Convert a diagram to a truth table Write the logic gate response for each gate and record the value of X.

Repeat this technique to fill in the remaining rows in the truth table. You now have the tools necessary to analyze older IC technology systems and convert them to an Altera CPLD IC design.

POWERPOINT PRESENTATION

Use PowerPoint to view the Lab 02 slide presentation.

LAB WORK PROCEDURE

VHDL Vending Machine System

To create a VHDL design, you must create a text file to enter the VHDL code. The text file is converted to a symbol that can be displayed in the Graphic Editor. The Graphic Editor is used to connect input and output symbols. Effectively, your VHDL code creates a new symbol for the **Altera symbol library**. Here is the procedure to create the text file for the VHDL code:

1. Turn on the computer (PC).
2. Insert a blank formatted floppy into drive A:. Use Windows Explorer to create a folder (or directory) called **Labs**. If you place VHDL designs in the root directory, you may get the following error: **I/O error: Can't open VHDL WORK**. Altera recommends that you maintain your VHDL design files in a subdirectory (or folder).
3. Start the **MAX+PLUS II** program. Create a project file. Click on the **File** menu and select **New**. You will see the "New" window. Refer to Figure 2-16. Select **Text Editor File** and click the "OK" button. You will see the Text Editor window inside the MAX+PLUS II window. Refer to Figure 2-17. Save the project. Click on "File," choose **Project** from the menu and then **Set Project To Current File**. You will see the "Save As" window. Refer to Figure 2-18. Make the changes shown in Figure 2-18 and then click the "OK" button. The titles at the top of the Text Editor window and the MAX+PLUS II window will show the title of the project: **vending**.
4. To reduce VHDL syntax errors (VHDL punctuation and grammar rules), Altera provides a number of VHDL templates. The templates reduce errors by providing a layout of a VHDL statement. Click on the **Templates** menu and select **VHDL Template....** The VHDL Template window will open. Refer to Figure 2-19.

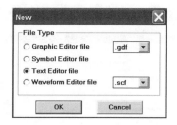

Figure 2-16 Altera "New" Window

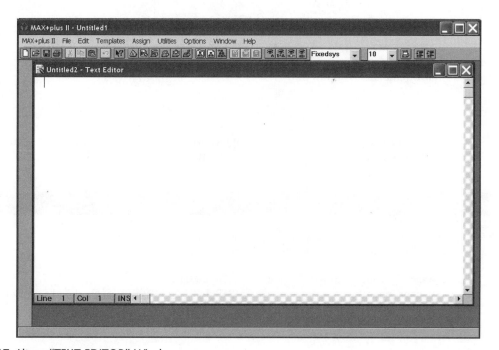

Figure 2-17 Altera "TEXT EDITOR" Window

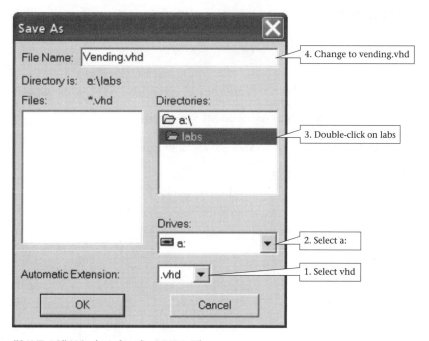

Figure 2-18 Altera "SAVE AS" Window for the VHDL File

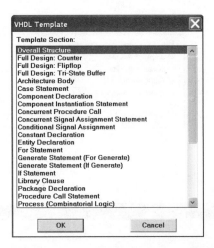

Figure 2-19 Altera "VHDL Template" Window

Select **Entity Declaration**. The template window will close and the text editor will display a generic entity declaration statement. Refer to Figure 2-20.

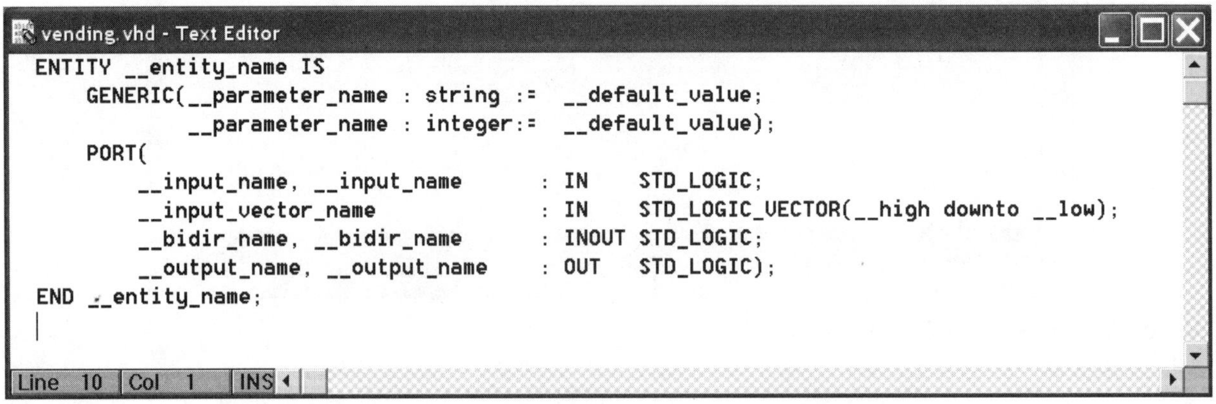

```
vending.vhd - Text Editor
ENTITY __entity_name IS
    GENERIC(__parameter_name : string :=  __default_value;
            __parameter_name : integer:=  __default_value);
    PORT(
        __input_name, __input_name     : IN    STD_LOGIC;
        __input_vector_name            : IN    STD_LOGIC_VECTOR(__high downto __low);
        __bidir_name, __bidir_name     : INOUT STD_LOGIC;
        __output_name, __output_name   : OUT   STD_LOGIC);
END __entity_name;

Line  10   Col   1     INS
```

Figure 2-20 Altera "TEXT EDITOR" Window with "VHDL Entity Declaration" Template

Reuse the **VHDL Template** window to insert the following templates:
    Architecture Body
    Concurrent Signal Assignment Statement
    Selected Signal Assignment Statement
Your Text Editor window should resemble Figure 2-21.

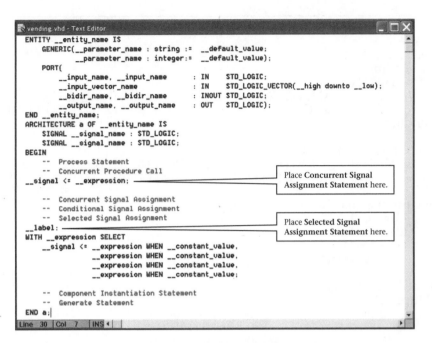

```
vending.vhd - Text Editor                                          _ □ ✕
ENTITY __entity_name IS
    GENERIC(__parameter_name : string :=  __default_value;
            __parameter_name : integer:=  __default_value);
    PORT(
         __input_name, __input_name        : IN    STD_LOGIC;
         __input_vector_name               : IN    STD_LOGIC_VECTOR(__high downto __low);
         __bidir_name, __bidir_name        : INOUT STD_LOGIC;
         __output_name, __output_name      : OUT   STD_LOGIC);
END __entity_name;
ARCHITECTURE a OF __entity_name IS
    SIGNAL __signal_name : STD_LOGIC;
    SIGNAL __signal_name : STD_LOGIC;
BEGIN
    --  Process Statement
    --  Concurrent Procedure Call
__signal <= __expression; ─────────────────┐   Place Concurrent Signal
                                            │   Assignment Statement here.
    --  Concurrent Signal Assignment
    --  Conditional Signal Assignment
    --  Selected Signal Assignment
__label: ──────────────────────────────────┐   Place Selected Signal
WITH __expression SELECT                    │   Assignment Statement here.
    __signal <= __expression WHEN __constant_value,
                __expression WHEN __constant_value,
                __expression WHEN __constant_value,
                __expression WHEN __constant_value;

    --  Component Instantiation Statement
    --  Generate Statement
END a;
Line  30  Col  7   INS ◄                                           ►
```

Figure 2-21 Altera "TEXT EDITOR" Window with All the VHDL Templates

5. VHDL templates often contain lines that are not required and use generic names such as __**entity_name**, __**input name**, etc. Change the VHDL code to create the final version of the code for the vending machine system. When you are done, your Text Editor window should resemble Figure 2-22. The text file that represents your VHDL code has been created. The text file needs to be converted into a symbol that can be displayed in the Graphic Editor.

6. Create a symbol for the VHDL vending machine. Click on the **File** menu and select **Create Default Symbol**. The "Compiler" window will open and run. This step is similar to the "Save & Check" step. If you get 0 errors and 0 warnings, you can close the compiler window and continue with the lab. If you have errors, skip to the section titled Correcting VHDL Syntax Errors in "Appendix B." Continue from this point when all errors have been corrected.

```vhdl
LIBRARY ieee;
USE ieee.STD_LOGIC_1164.ALL;
ENTITY vending IS
          PORT(
                    Q1, Q2, Q3, L                    : IN      STD_LOGIC;
                    P, C                             : OUT     STD_LOGIC);
END vending;

ARCHITECTURE a OF vending IS
          SIGNAL input: STD_LOGIC_VECTOR (3 DOWNTO 0);
          SIGNAL output: STD_LOGIC_VECTOR (1 DOWNTO 0);
BEGIN
-- Concurrent Signal Assignment
input     (3)     <= L;
input     (2)     <= Q3;
input     (1)     <= Q2;
input     (0)     <= Q1;


-- Selected Signal Assignment
WITH input SELECT
          output     <=          "00" WHEN "0000",
                                 "00" WHEN "0001",
                                 "00" WHEN "0010",
                                 "00" WHEN "0011",

                                 "00" WHEN "0100",
                                 "00" WHEN "0101",
                                 "00" WHEN "0110",
                                 "10" WHEN "0111",

                                 "11" WHEN "1000",
                                 "11" WHEN "1001",
                                 "11" WHEN "1010",
                                 "11" WHEN "1011",

                                 "11" WHEN "1100",
                                 "11" WHEN "1101",
                                 "11" WHEN "1110",
                                 "11" WHEN "1111",

                                 "00" WHEN others;

P         <=          output(1);
C         <=          output(0);
END a;
```

Figure 2-22 Altera "TEXT EDITOR" Window with Changes to the VHDL Templates

Use the VHDL Vending Machine in a Design

The text file that represents your VHDL code has been created. The text file has been converted into a symbol that can be displayed in the Graphic Editor. The Graphic Editor will be used to connect input and output symbols.

1. Create a project file. Click the File menu and select **New**. You will see the "New" window. Select **Graphic Editor File** and click the "OK" button. You will see the Graphic Editor window.
2. Save the project. Click the **File** menu, select **Project** and then **Set Project To Current File**. You will see the "Save As" window. Refer to Figure 2-23.

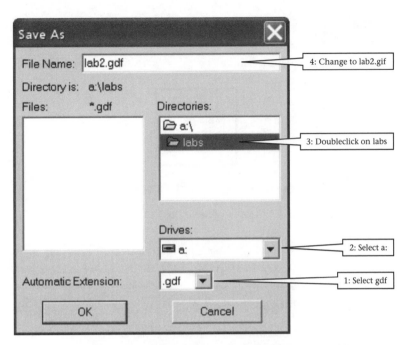

Figure 2-23 Altera "Save As" Window for The Graphic Design File (GDF)

Make the changes shown in Figure 2-23 and then click the "OK" button. The titles at the top of the Graphic Editor window and the MAX+PLUS II window will show the title of the project: **Lab2**.

3. Enter the symbol for the VHDL vending machine. With the **Selection Pointer** (pointer you can move with the mouse), double-click in an empty space in the Graphic Editor window. You will see the **ENTER SYMBOL** window. Refer to Figure 2-24. Double-click on the "**vending**" symbol in the **Symbol Files** window. The **VHDL vending machine symbol** is placed on the graphic designer worksheet. Refer to Figure 2-25.

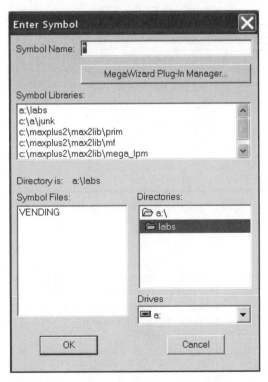

Figure 2-24 Altera "Enter Symbol" Window

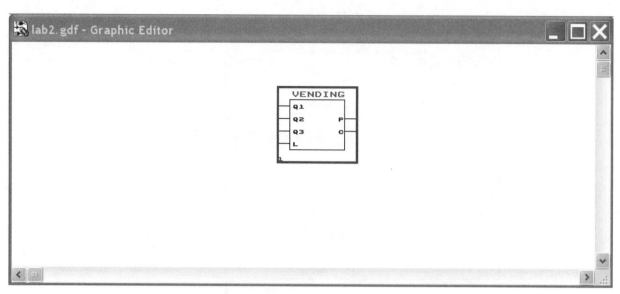

Figure 2-25 Altera "Graphic Editor": VHDL Vending Symbol

4. Complete the drawing for the VHDL vending machine. Use the Lab Work Procedure in Lab 1 as a guide. Complete steps 1 through 12. Be sure to ask the instructor which Altera CPLD you should use for the system design. You can design the system using either the MAX or the FLEX CPLD IC. The Graphic Editor file for the VHDL vending machine should resemble Figure 2-26.

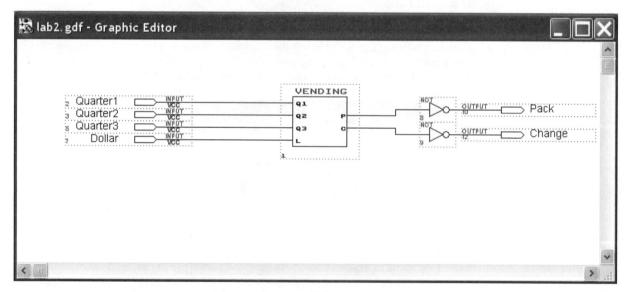

Figure 2-26 Altera "Graphic Editor": VHDL Vending Machine System

5. Lab 1 describes all the steps that are required to program a **UP** board with the vending machine system. Use the Lab Work Procedure in Lab 1 as a guide and complete steps 13 through 17. Connect inputs **Quarter1**, **Quarter2**, **Quarter3**, and **Dollar** to UP board switches. Connect "**Pack**" and "**Change**" to UP board LEDs.
6. At this point you have drawn the diagram, programmed the IC, and connected all switches and LEDs. The system is ready for testing.
   Test A. Flip all switches to the 0 position. This represents the condition where no money is inserted in the machine. Both LEDs should be *off*.

Test B. Flip only the three-quarter inputs to the 1 position. This test should result in the **Package** LED turning *on* and the **Change** LED turning *off*.

Test C. Flip only the **Dollar** input to the 1 position. This test should result in both the **Package** and **Change** LED turning *on*.

If your design seems to work properly, then you have completed the lab. If you need to make changes to the design, refer to the section titled "Making Changes to a Functional VHDL Design" in Appendix B. You may not have syntax errors but you may have entered some of the VHDL code incorrectly.

LAB EXERCISES AND QUESTIONS

This section contains lab exercises that can be performed on the UP board and questions that can be answered at home. Ask your instructor which exercises to perform and which questions to answer.

Lab Exercises

Exercise 1: Improve the Vending Machine System.

a. The vending machine system has flaws in the design. Here is a scenario that would frustrate a user of the vending machine: A user searches her pocket, finds a quarter, and inserts it into the machine. Then she realizes that she does not have two more quarters but that she has a dollar! If the user inserts the dollar, what would be the result? Test this condition on the vending machine system before answering the question.

b. Here is another scenario that would frustrate a user of the vending machine: A user searches her pocket, finds two quarters, and inserts them into the machine. Then she realizes that she does not have one more quarter, but that she has a dollar! If the user inserts the dollar, what would be the result? Test this condition on the vending machine system before answering the question.

c. To correct the problems described in a and b above, you can add additional change outputs ("Change2," "Change3"). Improve the vending machine by adding extra outputs to the design. Before testing the new system, some important points should be made regarding the UP board. Whenever you make a VDHL design change, you must make changes to the VDHL code, re-create the default symbol, and update the Graphic Editor Symbol. The section titled "Making Changes to a Functional VHDL Design" in Appendix B describes this process. Once the new default symbol is created and updated, then you must recheck for basic errors with the "Project Save & Check" procedure. You must regenerate a new report and programming files with the "Project Save & Compile" procedure. For a MAX IC design, you must also review the report file to verify the pin numbers. The compiler can reassign MAX pin numbers each time you recompile. You may need to move MAX switch and MAX LED wires. Finally, the MAX IC and FLEX IC must be reprogrammed. Test the system and record the results in a truth table with four inputs and four outputs.

Exercise 2: Design a digital sound meter.

A **sound converter** uses a microphone to monitor the intensity of sound. It generates a 4-bit number that represents the intensity of sound. The number 0 (binary 0000) represents absolute silence, and 15 (binary 1111) represents extreme noise. The sound converter sends the 4-bit number to a Digital Sound Meter. The Digital Sound Meter controls the *on/off* status of four LEDs. The LEDs will indicate the sound level. Refer to Figure 2-27.

VHDL Digital Sound Meter:

The sound converter system will not be part of the design. Four UP switches will be used to simulate the sound converter. Use variables S3, S2, S1, S0 to represent the 4-bit number from the sound converter. You will design the digital sound meter. The digital sound meter will be connected to four LEDs. Use variables R, Y, B, G to represent the status of the LEDs.

The LEDs will be controlled as follows:

The green LED "G" is ON when the meter receives 0, 1, or 2 from the converter.

The blue LED "B" is ON when the meter receives 3, 4, 5, or 6 from the converter.

The yellow LED "Y" is ON when the meter receives 7, 8, 9, 10, or 11 from the converter.

	Noise Level	Ear Protection
Red LED on:	Extremely noisy:	Earmuffs and earplugs required
Yellow LED on:	Noisy:	Earmuffs required
Blue LED on:	Low intensity noise:	Foam earplugs required
Green LED on:	Quiet:	No hearing protection required

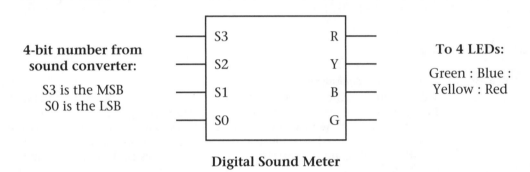

Figure 2-27 Sound Meter System for Exercise 2

The red LED "R" is ON when the meter receives 12, 13, 14, or 15 from the converter. Only one LED is lit at one time.

Create a truth table for the digital sound meter and use VHDL selected signal assignment to build and test the operation of the sound meter system on the Altera UP board.

Exercise 3: Design a drill machine system. (This exercise can be used as a project.)

A drill machine has a drill motor, a material clamp, two start buttons, and a protective cover. The protective cover can be raised above the drill machine or it can be lowered over the drill machine. Refer to Figure 2-28.

The user of the machine must place the material to be drilled into the clamp and secure it in place. The user can start the machine by pressing one of the two start buttons or by pressing both start buttons at the same time. One-button start-up or two-button start-up is dependent on the position of the protective cover (cover up or cover down). You will need to build and test the operation of the system to understand all of the operational details.

Inputs:

Start buttons: A, B. Button pressed down = logic 0
Cover sensor: C. Cover down = logic 0
Material clamp sensor: D. Part is clamped = logic 1

Output:

Drill Motor: X. Drill motor on = logic 1

Equation: $X = D \cdot [\overline{A} \cdot \overline{B} + \overline{C} \cdot (\overline{A \cdot B})]$

a. Convert the Boolean equation to a truth table and use VHDL selected signal assignment to build and test the operation of the drill machine on the Altera UP board.

b. Use the working UP system to answer the following questions:
  • How does the user start the drill machine with the cover up?
  • How does the user start the drill machine with the cover down?
  • The drill machine system ensures the safety of the user with two start buttons and a protective cover. Explain why the user is safe.

You may want to save the files for the drill machine and use them in Lab 3, Exercise 6. Ask your instructor for guidance regarding this matter.

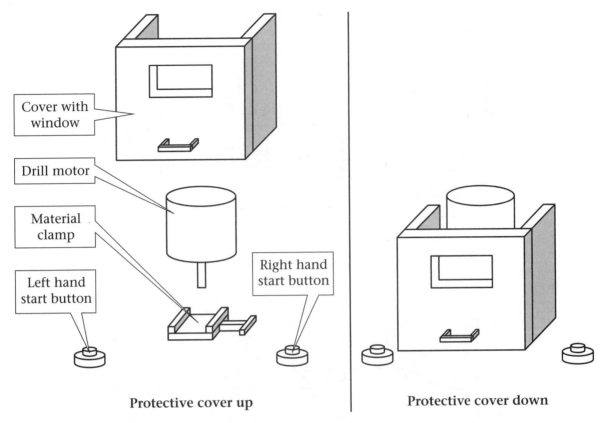

Protective cover up

Protective cover down

Figure 2-28 Drill Machine System for Exercise 3

Lab Questions

1. Substitute the binary values into the equation to generate a truth table.
   a. $X = \overline{A + B} \oplus \overline{C}$
   b. $X = \overline{\overline{A} \cdot B} + C$
2. Generate a truth table for the equation. Use the technique found in the introductory information section titled "Generate a Truth Table for Sections of the Equation."
   a. $X = A \cdot B + \overline{B \cdot C}$
   b. $X = \overline{A + B} \oplus \overline{C}$
3. Generate a truth table for each diagram. Refer to Figure 2-29.

Conclusions

- The VHDL "Selected Signal Assignment" allows you to enter a digital design as a truth table. This makes digital system design intuitive and easy to implement.
- The real world includes many digital systems implemented using older generation IC technology. Being able to analyze and convert these older systems to Altera CPLD IC technology is important.
- Before the advent of Altera CPLD ICs, digital systems design changes required a lot of time-consuming work. System changes take only minutes to implement using Altera programmable CPLD IC technology.
- To create a VHDL design, you must create a text file to enter the VHDL code. The text file is converted to a symbol that can be displayed in the Graphic Editor. The Graphic Editor is used to connect input and output symbols. Effectively, your VHDL code creates a new symbol in the Altera symbol library.

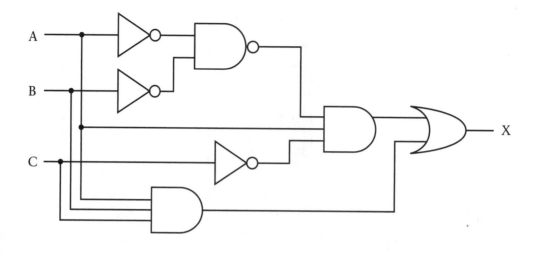

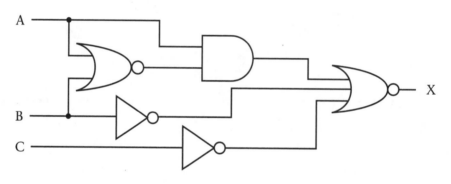

Figure 2-29 Logic System Diagrams for Question 3

- Syntax errors are like English grammar errors. Altera needs each line of VHDL code to have the correct punctuation. Always correct the first syntax error in the message list. The first syntax error is located in the line above the one indicated by the Altera compiler. This first error may be the cause of additional errors.

Flip-Flops, Shift Registers, and Switch Bounce

Equipment	Altera UP board
	Book CD-ROM
	Spool of 24 AWG wire
	Wire cutters
	Blank floppy disk.
Objectives	Upon completion of this lab, you should be able to:

- Operate SR and D flip-flops.
- Design and operate a shift register using D flip-flops.
- Debounce a switch.

INTRODUCTORY INFORMATION

In this lab you will learn how to use a flip-flop as a storage cell for binary data. Flip-flops are the fundamental building blocks of memory systems. You will also learn how switches bounce and how this bounce affects digital systems.

SR Flip-Flop Fundamentals

Flip-flops are digital devices that can store binary data. Refer to Figure 3-1. The Set (S) and Reset (R) inputs can be used to store a binary 1 or 0 at output Q. S and R can be used to flip Q high or flop Q low. The output $\overline{Q}$ is by definition always opposite to Q. In many systems $\overline{Q}$ is not used and is left unconnected, or it is omitted altogether. The Altera symbol library does not show this output.

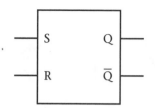

S	R	Q Response
0	0	No change mode. Hold mode.
0	1	Reset mode: Q = 0
1	0	Set mode: Q = 1
1	1	Ambiguous mode. Not allowed.

Figure 3-1 The SR Flip-Flop Behavior Table

The operation of a flip-flop is defined in a table called the behavior table. Figure 3-1 shows the behavior table. The S and R inputs are both active high. Active high means that *S = 1* sets the flip-flop (*S = 0* does not set). *R = 1* resets the flip-flop (*R = 0* does not reset). SET means make *Q = 1*. Reset means make *Q = 0*.

When *S = 0* and *R = 0*, then Q does not change. Q holds its logic level (1 or 0). It is equivalent to not issuing either the set or the reset command.

When *S = 1* and *R = 1*, then Q is ambiguous. Both Q and $\overline{Q}$ outputs become the same logic level. This breaks the definition of a flip-flop. You can think of it this way—*S = 1* says SET and *R = 1* says RESET. The flip-flop does not know whether to output a 0 or a 1 when both S and R are asserted. *S = R = 1* should never be used!

The NOR Gate SR Flip-Flop

Two cross-coupled NOR gates create a basic SR flip-flop called the NOR gate latch. Refer to Figure 3-2(A).

Figure 3-2(B) demonstrates the set mode. Each NOR gate is analyzed and reanalyzed until the Q output settles and no longer changes. The final logic level at the output confirms the set mode. The analysis can be repeated to demonstrate the **hold**, the **reset**, and the **ambiguous** mode.

Active Low SR Inputs

Two cross-coupled NAND gates create a basic flip-flop called the NAND gate latch. Using NAND gates makes the S and R inputs active low. Active low means that *S = 0* sets the flip-flop (*S = 1* does not set). *R = 0* resets the flip-flop (*R = 1* does not reset). Refer to Figure 3-3. When *S = 1* and *R = 1*, then Q does not change. Q holds its logic level (1 or 0). It is equivalent to not issuing either the set or the reset command. When *S = 0* and *R = 0*, then Q is ambiguous. Both inputs are asserted and the flip-flop does not know whether it should set or reset the output Q. *S = R = 0* should never be used!

Clocked SR Flip-Flops

Figure 3-4 shows a clocked SR flip-flop. The clock is a master control input. It must be asserted to make output Q respond to S and R inputs. It's like a key to a car. You need the key to drive the car.

The clock is a **positive edge triggered** input. That means a transition from 0 to 1 is required in order that S and R control output Q. Holding a constant logic 1 or a constant logic 0 at the clock input does not allow SR to change output Q.

A positive edge triggered clock input acts like a trigger on a starter's pistol. With a starter's pistol you must squeeze the trigger to fire the pistol. You must release the trigger and resqueeze it to fire the pistol a second time. Holding the trigger down will not fire the pistol a second time. An SR flip-flop needs a positive edge at the clock to trigger it into action. Holding the clock input high will not allow the flip-flop to change logic level. A new positive edge is required. Figure 3-5 shows the internal construction of a positive edge triggered SR flip-flop.

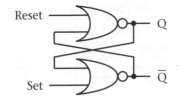

S	R	Q Response
0	0	No change mode. Hold mode.
0	1	Reset mode: Q = 0
1	0	Set mode: Q = 1
1	1	Ambiguous mode. Not allowed.

(A) NOR gate latch

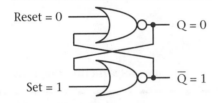

You must assume an initial condition at output Q. This is necessary because the outputs are connected to the inputs. This creates a feedback path that can only be analyzed when a starting point is assumed. Let's assume Q = 0.

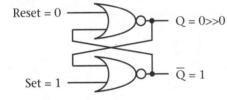

Place the logic levels at inputs Set and Reset. (S = 1 and R = 0)

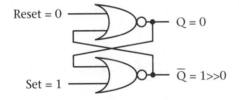

Analyze the top NOR gate and record Q. Q = 0 because 0,1 = 0

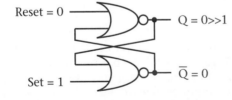

Analyze the bottom NOR gate and record $\overline{Q}$. $\overline{Q}$ = 0 because 0,1 = 0

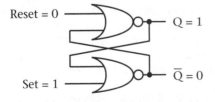

Reanalyze the top NOR gate and record Q. Q = 1 because 0,0 = 1.

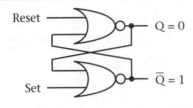

Reanalyze the bottom NOR gate and record $\overline{Q}$. $\overline{Q}$ = 0 because 1,1 = 0.

(B) The SET mode for the NOR gate latch

Figure 3-2 The NOR Gate Latch

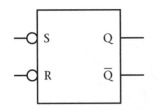

S	R	Q Response
0	0	Ambiguous mode. Not allowed.
0	1	Set mode: Q = 1
1	0	Reset mode: Q = 0
1	1	No change mode. Hold mode.

Figure 3-3 Active Low Input SR Flip-Flop

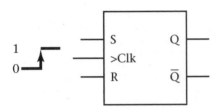

S	R	Q Response: On +ve Edge of Clock
0	0	No change mode. Hold mode.
0	1	Reset mode: Q = 0
1	0	Set mode: Q = 1
1	1	Ambiguous mode. Not allowed.

Figure 3-4 Positive Edge Triggered Flip-Flop

Internal structure diagram

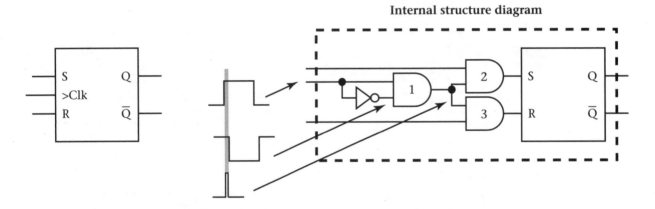

Note: This diagram is used to teach edge triggering concepts. It should not be programmed into a CPLD.

Figure 3-5 Inside the Positive Edge Triggered SR Flip-Flop

Figure 3-5 highlights the change of the clock input from 0 to 1. This change is a positive edge. At the instant the clock is 1, then AND gate #1 will have both inputs high. One input is high because the clock input is high, and the other input is high because the output of the NOT gate has not had time to respond to the clock input change from 0 to 1. With both inputs high, AND gate #1 outputs a high. AND gates #2 and #3 will transfer their logic levels to S and R. This condition, however, only lasts about 3 to 10 nanoseconds (10×10^{-9} seconds). That is the amount of time it takes for the NOT gate to respond to the clock and change its output from 1 to 0. The response time of the NOT gate delays its output from changing. This delay is called propagation delay. Propagation delay is the reaction time of the NOT gate. When the NOT gate outputs a 0, the flip-flop returns to the hold mode (S = R = 0). To restart this cycle the clock must return to logic 0 and then change back to logic 1.

Figure 3-6 shows a second variety of the clocked SR flip-flop that requires a 1 to 0 transition at the clock input. The symbol has a bubble at the clock input. It is called the negative edge triggered SR flip-flop.

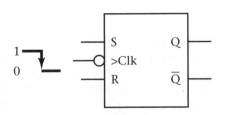

S	R	Q Response: On –ve Edge of Clock
0	0	No change mode. Hold mode.
0	1	Reset mode: Q = 0
1	0	Set mode: Q = 1
1	1	Ambiguous mode. Not allowed.

Figure 3-6 Negative Edge Triggered Flip-Flop

D Flip-Flop Fundamentals

The **D** flip-flop is a "Data" flip-flop. It is a flip-flop that can be used to store binary data or to shift binary data to other D flip-flops. Refer to Figure 3-7. The logic level at input D is transferred to output Q on the positive edge of the clock.

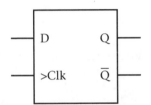

D	Q Response: On +ve Edge of Clock
0	Reset mode: Q = 0
1	Set mode: Q = 1

Figure 3-7 D Flip-Flop

Figure 3-8 shows the internal structure of a positive edge triggered D flip-flop. From Figure 3-8, you can see that when D = 1, the internal S = 1 and R = 0, which is the set mode. When D = 0, then the internal S = 0 and R = 1, which is the reset mode. There is also a negative edge triggered D flip-flop. Refer to Figure 3-9.

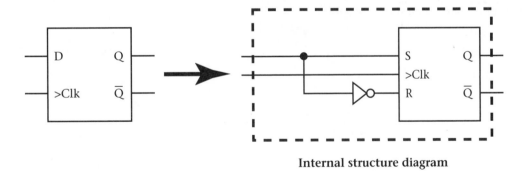

Internal structure diagram

Figure 3-8 D Flip-Flop Internal Structure

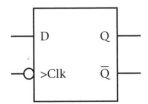

D	Q Response: On –ve Edge of Clock
0	Reset mode: Q = 0
1	Set mode: Q = 1

Figure 3-9 Negative Edge Triggered D Flip-Flop

Shift Register Fundamentals

Shift registers transfer a bit of binary data from one D flip-flop stage to the next. Refer to Figure 3-10(A). Figure 3-10(A) shows all D flip-flop stages sharing a common clock input. Asserting the clock provides a positive edge transition to all four flip-flops. This causes Din to transfer to Qa, Db to transfer to Qb, Dc to transfer to Qc, and Dd to transfer to Qd. The binary data from a D flip-flop on the left is transferred to the next flip-flop to the right.

Figure 3-10(B) demonstrates the response of the shift register to a single clock transition with Din = 1. The transition is a positive edge that asserts the operation of all the D flip-flops. It is referred to as a clock pulse.

Figure 3-10(C) demonstrates the response of the shift register to an additional three clock pulses. After a total of four clock pulses the shift register fills with the logic level from Din.

Using a shift register is analogous to using a bucket brigade to fill a huge container with water from a lake. A bucket brigade is a line of people with the first person in line standing in a lake. The last person in line stands next to a huge water container. The first person fills a bucket with water and passes a full bucket to the person next to him. The full bucket is passed from person to person to the last person in the line. The last person empties the bucket into the huge water container.

Switch Bounce

A switch is a mechanical device that closes contacts. **Switch Bounce** occurs when a switch is changed from logic 0 to logic 1 (or 1 to 0). The contacts chatter, causing a small electrical storm that lasts less than 20 milliseconds (20×10^{-3} seconds). Refer to Figure 3-11(A).

The logic gate systems constructed in Labs 1 and 2 have been able to absorb and dissipate switch bounce. When a switch was flipped, these systems became unstable for 20 milliseconds. After 20 milliseconds, these systems became stable and would output the correct response. The period of instability was so short that you could not even notice any negative effects caused by the switch bounce.

Switch bounce causes problems in digital systems that use flip-flops. Switch bounce generates several positive edge transitions and negative edge transitions during the 20-millisecond time interval. These additional transitions can clock a flip-flop system several times. Figure 3-11(B) shows the effect of switch bounce on a shift register system. A single flip of the switch is supposed to clock the shift register once. Unfortunately, when you flip the switch once, the system reacts as if you have flipped it five times because of the bounce. Switch bounce is not predictable. Sometimes a switch is flipped and it generates no switch bounce. The next time a switch is flipped it generates one, two, three, or more positive edges. If a switch is to be used to clock flip-flop stages, it must be de-bounced.

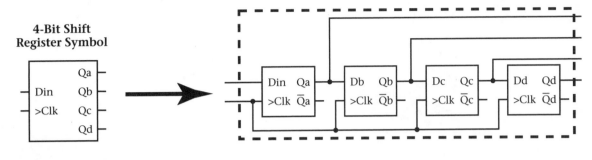

(A) 4-bit shift register

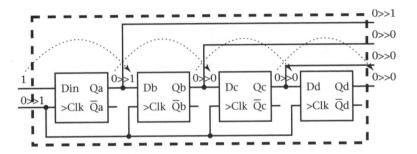

(B) Shift register response to a single clock transition with Din = L

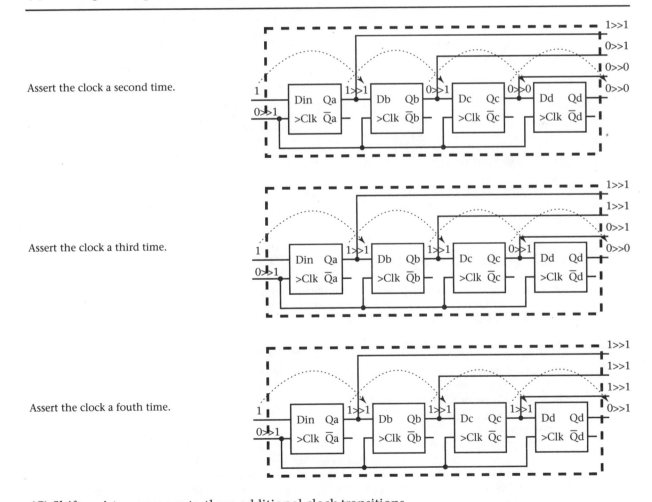

Assert the clock a second time.

Assert the clock a third time.

Assert the clock a fouth time.

(C) Shift register response to three additional clock transitions

Figure 3-10 4-Bit Shift Register Operation

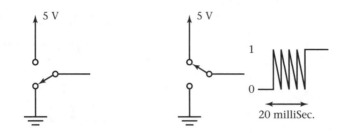

(A) Switch bounce

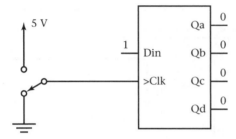

> You must assume an initial condition. Let's assume Qa = Qb = Qc = Qd = 0.

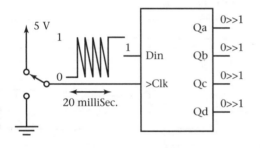

> Flip the switch ONCE. Switch bounce generates 5 positive edges. This will shift binary 1 into all shift register outputs. This is very frustrating to the user. Flip the switch once and the system reacts as if you have flipped it 5 times. Switch bounce is not predictable. Sometimes a switch is flipped and it generates no switch bounce. The next flip generates 2 positive edges. If a switch is to be used to clock flip-flop stages, it must be debounced.

(B) The effect of switch bounce on a shift register

Figure 3-11 Switch Bounce and Its Effect on Shift Register Systems

Switch Debounce System

Ironically enough, a switch debounce system uses a 4-bit shift register to eliminate the bounce. The system that was used to demonstrate the effects of switch bounce will now be used to eliminate it. Refer to Figure 3-12(A).

Debouncing a Switch That Is Flipped from 0 to 1

The top diagram in Figure 3-12(B) is the starting point of the analysis and shows the switch in the logic 0 position. The shift register is filled with 0s, and a new 0 flows into the shift register every 5 milliseconds. All Qs at 0 keep the AND gate output at 0. In this system the AND gate output represents the state of the switch. The switch is in the zero position and so is the AND gate output.

The middle diagram in Figure 3-12(B) shows that the switch has been flipped to the logic 1 position. The switch bounce causes Din to change between 1 and 0 several times over the 20-millisecond interval. The switch is bouncing while the shift register continues to be clocked every 5 milliseconds. The shift register can either capture a 1 if it is clocked while Din bounces high or it can capture a 0 if it is clocked while Din bounces low. The chance of either of these events occurring is even. If a 0 is captured, then the AND gate output will stay at 0 and the switch assertion goes undetected. If a 1 is captured, then the AND gate output will also stay at 0 because

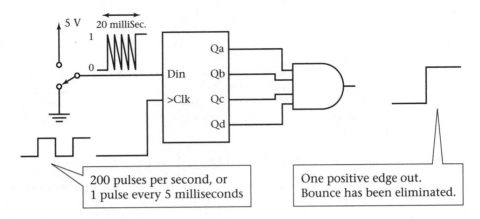

200 pulses per second, or
1 pulse every 5 milliseconds

One positive edge out.
Bounce has been eliminated.

(A) Switch de-bounce system

Start with the switch in the logic 0 position:

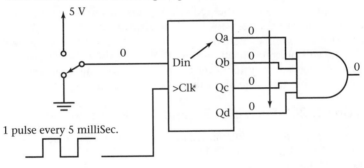

Flip the switch to the logic 1 position:

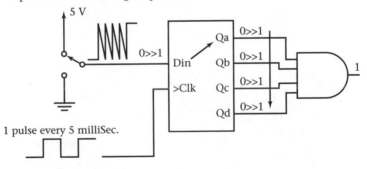

Flip the switch to the logic 0 position:

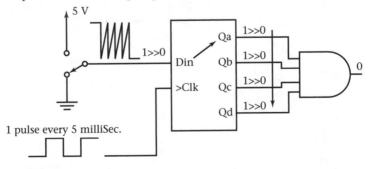

(B) Switch de-bounce system responding to a switch change from 0 to 1 and back to 0

Figure 3-12 Switch De-bounce System

the AND gate needs all four Q outputs to be 1 before it changes to 1. Once again the switch assertion goes undetected. The switch assertion will continue to be undetected until the switch stops bouncing and the shift register has a chance to shift in four consecutive 1s. When this occurs, 20 milliseconds will have elapsed and the AND gate output will change to 1. Once again the AND gate output will match the state of the switch. It also means that by the time the shift register can be filled with 1s, the bounce will have dissipated and there is no longer any chance of a 0 being shifted into the shift register. The AND gate will hold its output at 1 while a continuous flow of 1s is shifted from the stable switch into the de-bounce system every 5 milliseconds. The final result is one clean positive edge at the output of the AND gate and the AND gate output matches the switch position.

Debouncing a Switch That Is Flipped from 1 to 0

The bottom diagram in Figure 3-12(B) shows the switch has been flipped back to the logic 0 position. The switch bounce causes Din to change between 0 and 1 several times over the 20-millisecond interval. The switch is bouncing while the shift register continues to be clocked every 5 milliseconds. The shift register can either capture a 1 if it is clocked while Din bounces high or it can capture a 0 if it is clocked while Din bounces low. The chance of either of these events occurring is even. If a 1 is captured, then the AND gate output stays at 1 and the switch assertion will continue to be undetected. Eventually, a 0 is captured, which will immediately change the AND gate output to 0 because the AND gate needs all four Q outputs to stay at 1. The only way that the AND gate can change back to 1 is to shift in four consecutive 1s. This cannot occur because the switch bounce dissipates within 20 milliseconds and the switch will settle down to 0. The AND gate will hold its output at 0 while a continuous flow of 0s is shifted from the stable switch into the debounce system every 5 milliseconds. The final result is one clean negative edge at the output of the AND gate and the AND gate output matches the switch position.

Generating A 200-Pulse-per-Second Clock Signal Using the Altera UP Board

As you have seen, the switch debounce system needs a 200-pulse-per-second (PPS) clock signal. The UP board has a built-in 25.175 MHz oscillator. 25.175 MHz = 25.175 megahertz = 25,175,000 PPS. Refer to Figure 3-13.

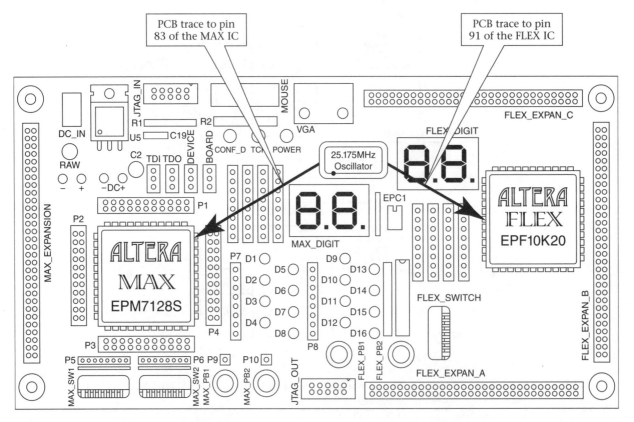

Figure 3-13 UP Board Oscillator

Unfortunately, the speed of the oscillator (25 million PPS) is too fast for the switch debounce system. A digital device called a **counter** can be used to slow down the oscillator PPS rate. Counter theory is explained in detail in Lab 5; however, counters can be used now if you understand their basic theory. The basic theory can be summarized with this statement: "Each successive counter output divides the pulse rate in half."

Figure 3-14 shows how a 25-stage counter can be used to slow down the PPS rate. The pulse rate can be calculated as follows:

Q0 = 1/2 of 25.175 MHz : Q1 = 1/2 of Q0: Q2 = 1/2 of Q1: ...   Q25 = 1/2 of Q24:

Or use this formula:

$$\text{Qn rate} = \frac{25{,}175{,}000}{2^{n+1}}$$

where n = output number

Output Q16 at 192 PPS is the closest to the value to 200 PPS that is required by the debounce system. The 192 PPS signal can be used in the debounce system without causing any problems.

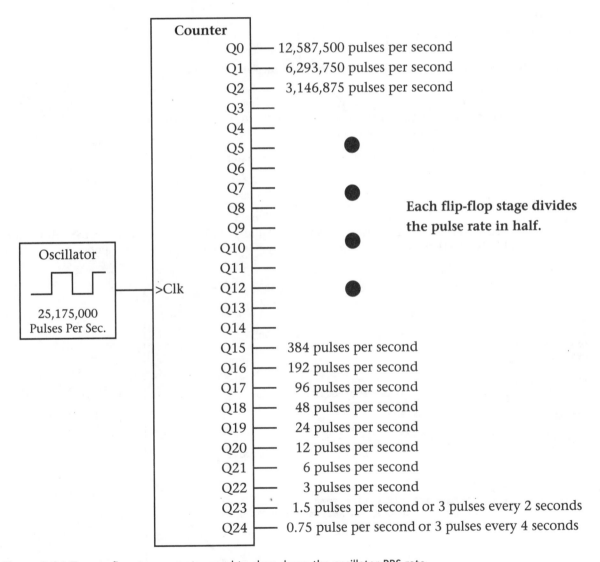

Figure 3-14 Twenty-five stage counter used to slow down the oscillator PPS rate

The Altera component library has a counter symbol that allows the creation of a 32-stage counter. The 32 output wires are drawn as a group called a bus line. The bus line is a thick line that represents 32 outputs. Refer to Figure 3-15.

A connection to a bus line is done by **NAME**. This means you name the bus line and then assign the same name to a single line to make a connection. The single line is logically connected

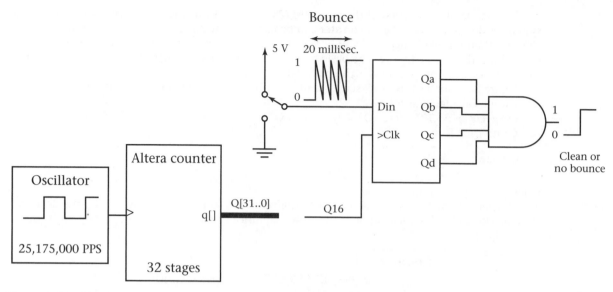

Figure 3-15 Thirty-two stage counter used to slow down the oscillator PPS rate

to the bus. This system connects the debounce system to Q16 of the counter.
The pulse rate at Q16 is calculated as follows:

$$\text{Q16 rate} = \frac{25,175,000}{2^{16+1}} = 192.07 \text{ PPS (approximately 200 PPS)}$$

VHDL Shift Register

The VHDL code for a shift register is shown in Figure 3-16. The statements **PROCESS (clk)** and **IF (clk'EVENT AND clk = '1')** will respond to every positive edge at the clock input.

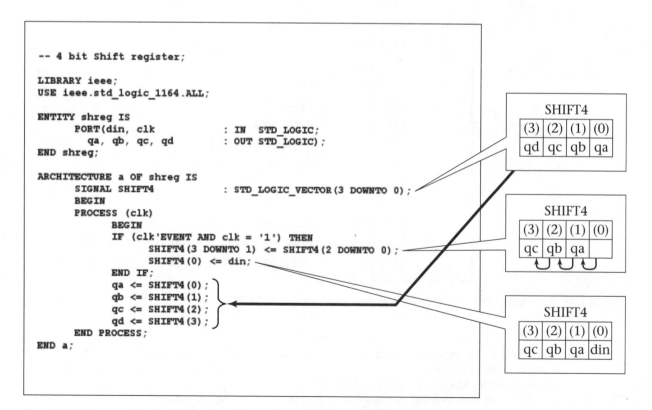

Figure 3-16 VHDL Shift Register

The response to a positive edge is:
    SHIFT4(3 DOWNTO 1) <= SHIFT4(2 DOWNTO 0);
    SHIFT4(0) <= din;
This is equivalent to:
    SHIFT4(3) <= SHIFT4(2) ;
    SHIFT4(2) <= SHIFT4(1) ;
    SHIFT4(1) <= SHIFT4(0) ;
    SHIFT4(0) <= din ;
These concurrent statements shift the information from one flip-flop stage to the next.

UP Board Pushbuttons

The UP board includes pushbutton switches. Refer to Figure 1-26. Pushbuttons are similar to buttons you find on a telephone. You press and hold them down for logic 0. You release them and they spring back up to logic 1. Unlike the switches you have been using up until now these switches can only maintain a 1. Once debounced, these switches are ideally suited for being used as the clock input to flip-flop systems like shift registers. Press the switch down to generate a negative edge and release it to generate a positive edge. One of these two edges will clock either a positive or a negative edge triggered shift register and the system is ready for the next clock signal.

MAX Pushbuttons

The MAX IC pushbuttons have wire connection sockets labeled P9 and P10.

FLEX Pushbuttons

The FLEX IC has two onboard pushbuttons. PCB traces connect the two switches to the FLEX IC. You do not need to connect external wires from the switches to the IC. You do need to assign pin numbers to the FLEX pushbuttons. FLEX pushbutton 1 is connected to **pin 28**. FLEX pushbutton 2 is connected to **pin 29**.

POWERPOINT PRESENTATION

Use PowerPoint to view the Lab 03 slide presentation.

LAB WORK PROCEDURE

Build and Test a Switch Debounce System

You will build a **switch debounce system** and place it in the symbol library so that it can be accessed in any future lab that needs a debounced switch. You will then build a **4-bit shift register system** and test its operation with and without the debounce system attached to the Altera pushbutton switch. You will gain first-hand knowledge of switch bounce and its effects on a flip-flop system.

Switch Debounce System

The first step is to create a library symbol for the debounce system that can be accessed today and at any other time in the future.
1. You will draw a diagram for a switch debounce system. Use the Lab Work Procedure in Lab 1 as a guide, and complete steps 1 through 10. Create a folder (directory) named **Labs**. Use the project name **debounce**. The symbol for a D flip-flop in the "prim" library is **DFF**. The symbol for the four-input AND in the "prim" library is **AND4**. The Graphic Editor file for the debounce system should resemble Figure 3-17.
2. A 200 PPS clock needs to be connected to the clock input of the shift register. It will clock the shift register once every 5 milliseconds. A counter can be used to slow down the UP board oscillator clock rate from 25,175,000 PPS to about 200 PPS (1 pulse every 5 milliseconds). Enter the symbol for the **LPM_COUNTER**. With the selection pointer

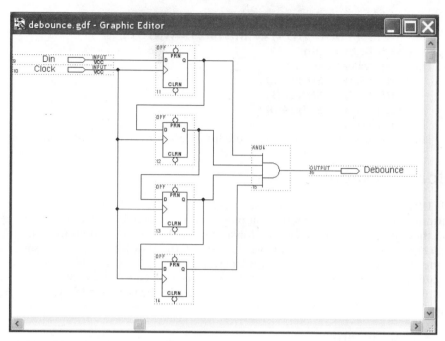

Figure 3-17 Altera "Graphic Editor": A Switch Debounce System

(pointer you can move with the mouse), double-click in an empty space to the left of the shift register in the Graphic Editor window. You will see the **Enter Symbol** window. Double-click on the **mega_lpm** library. Locate and double-click on the **LPM_COUNTER** symbol from the menu. The symbol will be placed on the Graphic Editor worksheet, and the "Edit/Ports Parameter" window will appear. Refer to Figure 3-18.

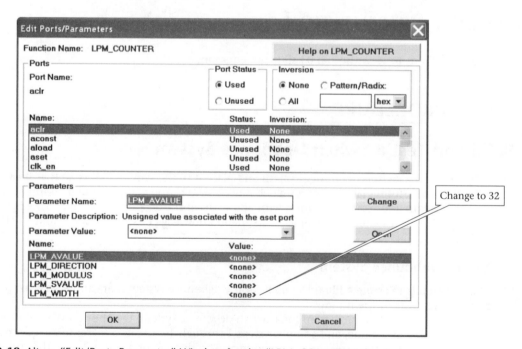

Figure 3-18 Altera "Edit/Ports Parameter" Window for the "LPM_COUNTER" Symbol

Change the value of "LPM_WIDTH" to **32** (see Figure 3-18) and click on the "OK" button. The symbol will now appear on the Graphic Editor worksheet. Refer to Figure 3-19.

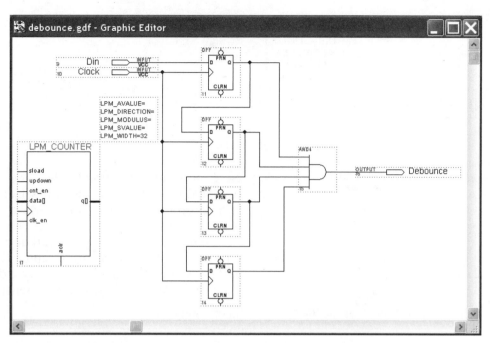

Figure 3-19 Altera "Graphic Editor": The Switch Debounce System with Counter

3. Reconnect the clock input symbol to the input of the **LPM_COUNTER** symbol. Refer to Figure 3-20.

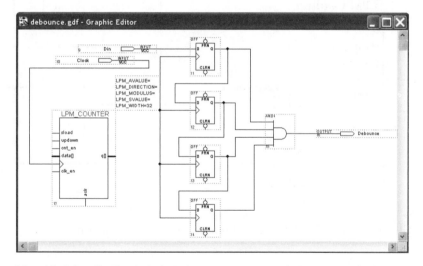

Figure 3-20 Altera "Graphic Editor": The Switch Debounce System with Counter and Clock Input

4. Draw a bus line coming out of the **LPM_COUNTER** output **q[]**. To place the **NAME Q[31..0]** on the bus line, you must click the **selection pointer** on the bus line. The bus line turns red and a small square insertion point appears below the bus line to show the insertion point of the name. Type **Q[31..0]**. Refer to Figure 3-21.
5. Output "Q16" from the "LPM_COUNTER" has a pulse at a rate of 192 PPS. This pulse rate is close enough to the 200 PPS required by the switch debounce system. Draw a line to the D flip-flop clock inputs and name the line **Q16**. To place the NAME Q16 on the line, you must click the selection pointer on the line. The line turns red and a small square insertion point appears below the line to show the insertion point of the name. Type **Q16**. Refer to Figure 3-22.

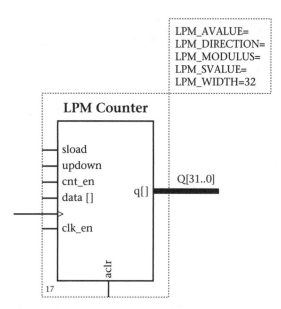

Note: It is important that the bus line turns red when you first click the selection pointer on the bus line. If the bus line does not turn red then the bus has not been named. It is also important that you type in exactly Q[31..0] with a capital Q and the square brackets with exactly 2 dots between 31 and 0. Altera is syntax sensitive.

Figure 3-21 Altera "Graphic Editor": Name the "LPM_Counter" Output Bus

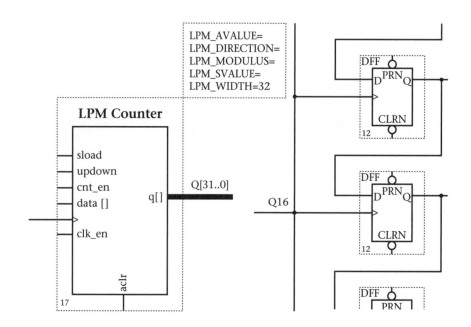

Figure 3-22 Altera "Graphic Editor": Connect the shift register clock to the LPM_Counter.

6. The debounce system drawing is complete. You need to convert the drawing into a library symbol. The symbol can be made to appear in the "Enter Symbol" window. You must click on the **File** menu and select **Create Default Symbol**. Visually, nothing appears to happen, but behind the scenes the symbol will be placed in the symbol library.

4-Bit Shift Register System:

You will attach the debounce system to a 4-bit shift register system. This will create a system that will allow you to test a 4-bit shift register as well as test the effect of switch bounce on a shift register.

1. You will draw a diagram for a **4-bit shift register**. Use the Lab Work Procedure in Lab 1 as a guide and complete steps 1 through 12. Use the project name **Lab3**. Save all files in

a folder (directory) named **Labs**. The symbol for a D flip-flop in the "prim" library is **DFF**. The debounce symbol will appear in the "Enter Symbol" window because of the work completed in the previous section of the lab. The Graphic Editor file for the 4-bit shift register system should resemble Figure 3-23.

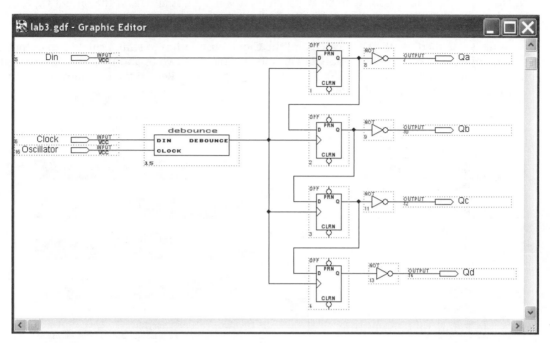

Figure 3-23 4-Bit Shift Register System

2. Lab 1 describes all the steps that are required to program a UP board with the 4-bit shift register. Use the Lab Work Procedure in Lab 1 as a guide and complete steps 13 through 17. Ignore any compiler warning and information messages.

MAX and FLEX IC Considerations

MAX IC Design:
- Connect input "Din" to a UP board MAX switch.
- Connect input "Clock" to a UP board MAX pushbutton.
- Assign input **Oscillator** to pin number **83**. A PCB trace connects the 25,175,000 PPS oscillator to pin 83 of the MAX IC. Refer to Lab 1, step 13$_{FLEX}$ for assistance regarding assigning an input symbol to a pin number.
- Connect outputs Qa, Qb, Qc, and Qd to UP board MAX LEDs.

FLEX IC Design:
- Assign input "Din" to a UP board FLEX switch.
- Assign input **Clock** to pin **28** or pin **29** to connect to a FLEX pushbutton.
- Assign input **Oscillator** to pin number **91**. A PCB trace connects the 25,175,000 PPS oscillator to pin 91 of the FLEX IC.
- Assign outputs Qa, Qb, Qc, and Qd to UP board FLEX LEDs. Remember to turn off unused LEDs by connecting them to Vcc.

3. At this point you have drawn the diagram, programmed the IC, and connected or assigned all switches and LEDs. The 4-bit shift register system is ready for testing.

LAB EXERCISES AND QUESTIONS

This section contains lab exercises that can be performed on the UP board and questions that can be answered at home. Ask your instructor which exercises to perform and which questions to answer.

Lab Exercises

Exercise #1: Test the shift register system.

When the power is first applied to the UP board, the initial conditions at output **Qa**, **Qb**, **Qc**, and **Qd** of the UP board are likely to be all 1s, all 0s, or a mix of 1s and 0s. If the output **Qa** is initially a 1, then use the **Din** switch and the **Clock** pushbutton to shift a single 0 to output **Qa**. Use the clock pushbutton to move the 0 to the other outputs **Qb**, **Qc**, and **Qd**. If the output **Qa** is initially a 0, then use the **Din** switch and the **Clock** pushbutton to shift a single 1 to output **Qa** then move it to the other outputs **Qb**, **Qc**, and **Qd**. Try shift two 0s and then two 1s. Try shifting a sequence of alternating 1s and 0s through the shift register. Record the last transfer in a chart similar to the one shown in Figure 3-24.

	Clock Pulse	Din	Qd	Qc	Qb	Qa
	Initial Conditions >>>>					
1	Before pressing button					
	Press and release button					
2	Before pressing button					
	Press and release button					
3	Before pressing button					
	Press and release button					
4	Before pressing button					
	Press and release button					
5	Before pressing button					
	Press and release button					

Figure 3-24 Shift Register Test Results

Exercise 2: Test the shift register system without the debounce symbol.

You will temporarily bypass the debounce symbol to test the effect of switch bounce on a shift register system. Figure 3-25 shows the changes you must make to bypass the debounce symbol.

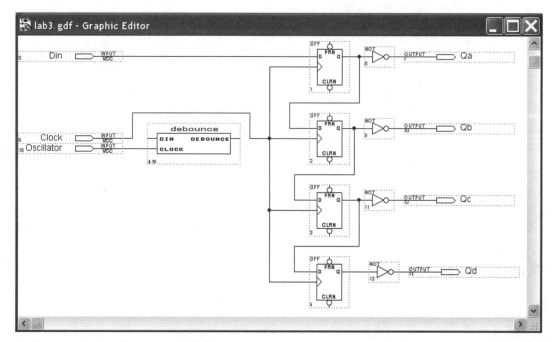

Figure 3-25 Bypassing the Debounce Symbol

Recompile and reprogram the Altera IC. Ignore any compiler warning messages. Try to repeat the tests described in Exercise 1.

   a. Explain how the system functions.

   b. About how many times does the switch bounce clock the shift register erroneously? Does a switch always produce the same number of bounce transitions?

   c. Try pressing the pushbutton various ways. Tap it and release it quickly. Press it down and hold it down. Does the way you tap the button seem to have any effect on switch bounce?

Restore the system to its original working state. Reconnect the debounce symbol.

Exercise 3: Test a shift register system that recirculates output Qd to input Qa.

You will disconnect input **Din**, add a **NOT** gate, connect the **NOT** gate output to **D**, and connect the **NOT** gate input to **Qd**. Figure 3-26 shows the changes you must make. Recompile and reprogram the Altera IC. Ignore any compiler warning messages. Test the operation of the new system and record its operation in a chart similar to the one shown in Figure 3-27. Explain how the system functions. How does the NOT gate control the response of the shift register?

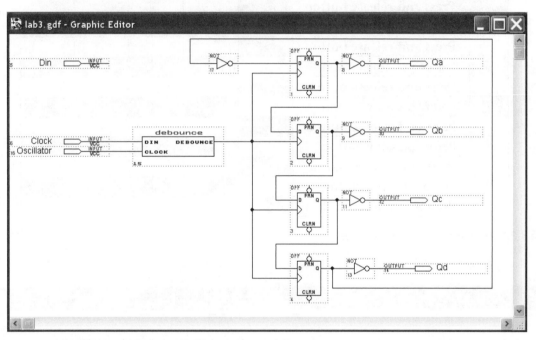

Figure 3-26 Shift Register System with Qd to Qa Recirculation

Exercise 4: Use the Altera simulator to test the operation of the recirculating shift register.

(This exercise can be done at home.)

In Exercise 3 you tested the operation of a recirculating shift register. You will use the Altera simulator to repeat the test. The output NOT gates and the debounce symbol have been removed from the diagram in order to simplify the system. Figure 3-28 shows the changes.

	Clock Pulse	D	Qd	Qc	Qb	Qa
	Initial Conditions >>>>					
1	Before pressing button					
	Press and release button					
2	Before pressing button					
	Press and release button					
3	Before pressing button					
	Press and release button					
4	Before pressing button					
	Press and release button					
5	Before pressing button					
	Press and release button					
6	Before pressing button					
	Press and release button					
7	Before pressing button					
	Press and release button					
8	Before pressing button					
	Press and release button					
9	Before pressing button					
	Press and release button					

NOT output

Figure 3-27 Recirculating Shift Register Test Results

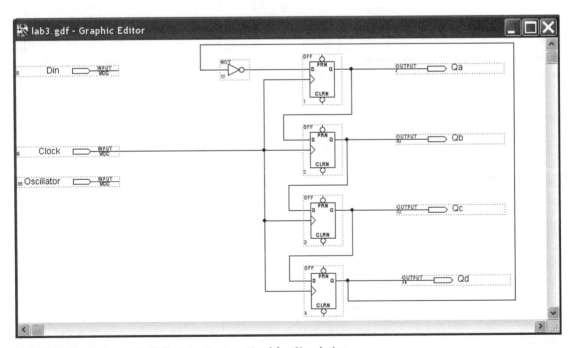

Figure 3-28 Recirculating a Shift Register System Used for Simulation

Use Appendix F as a guide to run a simulation of the recirculating register you have just created. The changes may give warnings when you use "Save, Compile & Simulate." These warnings can be ignored. The "Waveform Editor" window for the simulation should resemble Figure 3-29.

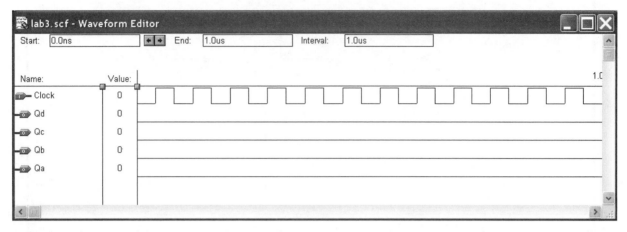

Figure 3-29 Altera "Waveform Editor": Recirculating the Shift Register System

Set the **End time** to 1.0us. The "Clock" uses an "Overwrite clock period" = 80.0ns. If the "Clock Period" entry box is grayed out (not selectable), then you should exit the "Overwrite Clock" window, select the **Options** menu, and remove the check mark beside the menu item "Snap to Grid." Return to the "Overwrite Clock" window and you will be able to set the clock period.
   a. Run the simulation and explain the output response you observe at the "Q" outputs.
   b. Group the "Q" outputs into a single node and display the group as a binary number. Run the simulation again. You may need to increase the scale (use the magnify glass with a "+" inside) to view the response of the simulation. Explain the output response you observe at the grouped node for output "Q."

Exercise 5: Convert the 4-bit shift register to VHDL.
Change the 4-bit register to a VHDL shift register. Refer to Figure 3-30. The VHDL code for a shift register is found in the section "VHDL Shift Register" in the Introductory Information. The code is also on the CD-ROM that accompanies the book. Use Lab 2 as a guide to design the VHDL shift register. Use the procedure shown in Lab Exercise 1 to test the shift register system.

Exercise 6: Add pushbuttons and an SR flip-flop to the drill machine system of Lab 2.
(This exercise can be assigned as a project.)
   The drill machine system in Lab 2: Exercise 3 can be improved by adding two pushbuttons to start the drill. Press and hold the start pushbuttons down to start the drill. Unfortunately, the drill machine stops when the pushbuttons are released. The drill machine operation can be improved even more by adding an SR flip-flop to maintain the drill in the *on* state. The SR flip-flop can be used as a memory cell to keep the drill machine running even after the pushbuttons are released. Your job is to add these improvements to the basic drill machine system of Lab 2.
   Here is a general guide that you can use to add these improvements:
   a. Use the drill machine disk files from Lab 2: Exercise 3. Connect (or assign) the UP board pushbuttons to "Start" button "A" and "B."
   b. Add a new switch input. It is a **kill** switch that will be used to turn off the drill machine.
   c. Add an SR flip-flop (NOR latch) to the drill output. The symbol in the mf library is **norltch**.
   d. Figure out how to connect the SR flip-flop (NOR latch) inputs to the drill machine system. You will need additional logic gates. Here is functional description of how the drill should work:
      • The SR flip-flop (NOR latch) will keep the drill running even after the pushbuttons are released.
      • The kill switch will stop the drill machine.
      • The material clamp sensor (D) will stop the drill machine if material should become unclamped.

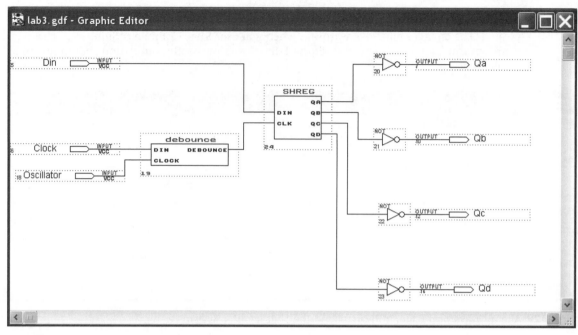

Figure 3-30 Altera "Graphic Editor": VHDL Shift Register System

   • At no time should the flip-flop be placed in the ambiguous mode.
   e. Test the operation of the system. Make sure the system starts and stops the drill
      according to all conditions described. Make sure that you *can't* place the flip-flop in the
      ambiguous mode.

Lab Questions

1. a. Draw the output waveforms for Q1 and Q2. Initial conditions: Q1 = Q2 = 0. Refer to
      Figure 3-31.
   b. D flip-flops can be used to delay data. Look at the waveforms and explain what the
      delay is.
2. Complete the table for each 4-bit shift register shown in Figure 3-32.

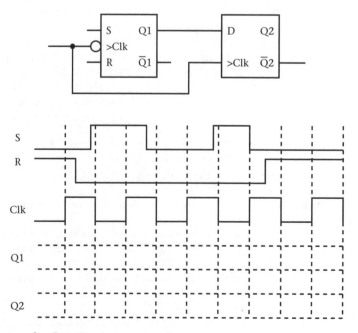

Figure 3-31 Flip-Flop System for Question 1

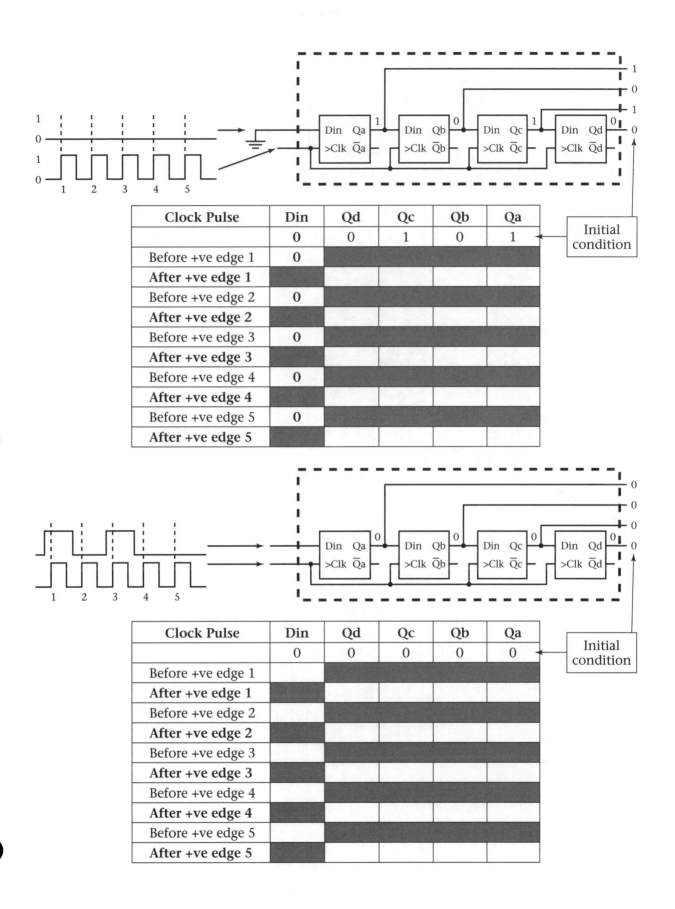

Figure 3-32 Shift Register Systems for Question 2

3. Complete the table for the 4-bit shift register shown in Figure 3-33. Fill in the column Din, Qa, Qb, Qc, and Qd. Leave the *AND Out* column open.

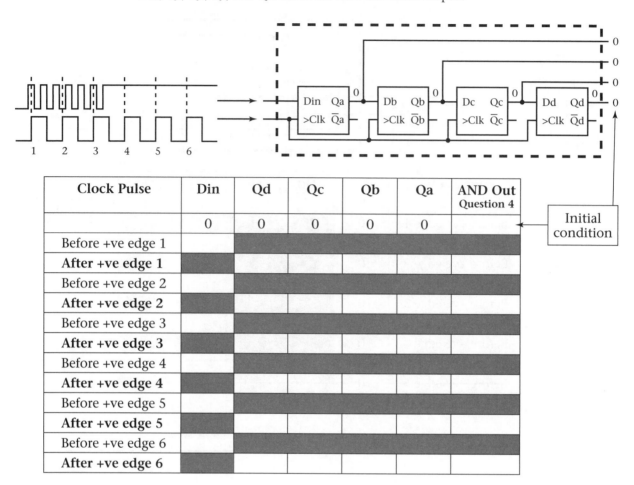

Clock Pulse	Din	Qd	Qc	Qb	Qa	AND Out Question 4
	0	0	0	0	0	
Before +ve edge 1						
After +ve edge 1						
Before +ve edge 2						
After +ve edge 2						
Before +ve edge 3						
After +ve edge 3						
Before +ve edge 4						
After +ve edge 4						
Before +ve edge 5						
After +ve edge 5						
Before +ve edge 6						
After +ve edge 6						

Figure 3-33 Shift Register System for Question 3

4. The diagram shown in Figure 3-34 can be used to demonstrate the operation of a switch debounce system.

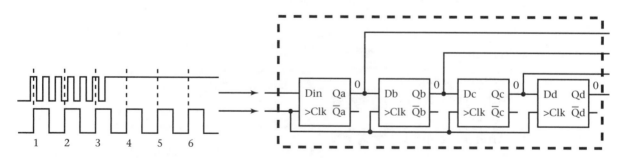

Figure 3-34 Shift Register System for Question 4

To make this into a debounce system you need to:
• Add an AND gate to the diagram.
• Label the pulse waveform that generates 1 pulse every 5 milliseconds from the UP board oscillator.

      • Label the pulse waveform that represents switch bounce.
      a. Fill in the *AND out* column in the table of question 3.
      b. Explain how the system eliminates switch bounce.
   5. Write the VHDL code for an 8-bit shift register.

CONCLUSIONS

- SR flip-flops can be used to store binary information.
- A clock input is a master control. It must be asserted to make output Q respond to inputs S and R.
- An edge-triggered clock is like the trigger of a starter's pistol. The edge must be regenerated in order to retrigger the clock input. It is like releasing and repulling the trigger of a starter's pistol.
- A D flip-flop is a data flip-flop. Q = D when the clock is asserted.
- D flip-flops can be chained together to create a shift register system. A shift register is like a bucket brigade. It passes data from one stage to the next whenever it is clocked.
- Switch bounce is an electrical storm that generates several clock edge signals in a 20-millisecond time interval. Switch bounce can cause erroneous clocking of flip-flops.
- A shift register system can be used to eliminate switch bounce.

  NOTE: Upcoming labs will need the debounce system. Do not delete files that begin with file name "debounce" in folder "Labs." Store this disk safely until it's needed.

Lab 4

Serial and Parallel Data Transfer Systems

Equipment   Altera UP board
Book CD-ROM
Spool of 24 AWG wire
Wire cutters
Blank floppy disk

Objectives   Upon completion of this lab, you should be able to:

- Design and operate a serial transfer system.
- Build and operate a data register.
- Design and operate a parallel transfer system.

INTRODUCTORY INFORMATION

In Lab 3, D flip-flops were used to store binary data, shift binary data, and debounce a switch. In this lab you will learn how to load numbers into a shift register and then use it to transmit binary data between two separate digital systems.

Flip-Flop Asynchronous Inputs

All flip-flops have two additional inputs called **Asynchronous Inputs** (**Pr** and **Clr**). They allow the user to set initial conditions at Q. Refer to Figure 4-1.

$\overline{Pr}$ is called the (not) preset input. Connecting it to 0 sets Q to 1. It is active low.

$\overline{Clr}$ is called the (not) clear input. Connecting it to 0 resets Q to 0. It is active low.

$\overline{Pr}$ and $\overline{Clr}$ are called **asynchronous** because they can be used to set and reset the flip-flop without the need of a clock signal. **D** and **Clk** are called **synchronous** inputs. The clock signal must be asserted to have D change Q.

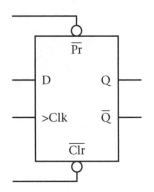

$\overline{Pr}$	$\overline{Clr}$	Response
0	0	AMBIGUOUS
0	1	Q = 1
1	0	Q = 0
1	1	D and Clk control Q (synchronous)

Figure 4-1 D Flip-Flop with Asynchronous Inputs

$\overline{Pr}$ and $\overline{Clr}$ override the conditions at the synchronous inputs (D and Clk) if they are left asserted (connected to 0). $\overline{Pr} = \overline{Clr} = 1$ allows D and Clk to control Q. $\overline{Pr} = \overline{Clr} = 0$ is not allowed, or AMBIGUOUS.

Using Asynchronous Inputs to Load a Shift Register

Figure 4-2 shows a 4-bit shift register with the asynchronous inputs configured to load in the number 10. Inputs Load, A, B, C, D can be used to load a 4-bit number into the shift register prior to using the clock to shift the data. Here is how it works:

Step 1: Place the number 10 at inputs DCBA. (Note: 10 = 1010 in binary.)

Refer to Figure 4-2(A).

Step 2: Connect Load to 0.

The NAND gates and inputs Load, D, C, B, A work together to set and reset the flip-flops to get the number 10 loaded into shift register outputs Qd, Qc, Qb, Qa. Remember, $\overline{Pr} = 0$ to preset Q and $\overline{Clr} = 0$ to reset Q. Refer to Figure 4-2(B).

Step 3: Connect Load to 1.

Load = 1 will make all NAND gates output 1. This ensures $\overline{Pr} = \overline{Clr} = 1$. This allows the synchronous inputs (D and Clk) to control Q. Remember, $\overline{Pr}$ and $\overline{Clr}$ override Q if they are left asserted (0). The Clock can now be used to shift data from Qa to Qb to Qc to Qd. Inputs A, B, C, D have no effect on Qa, Qb, Qc, Qd when Load = 1. Refer to Figure 4-2(C).

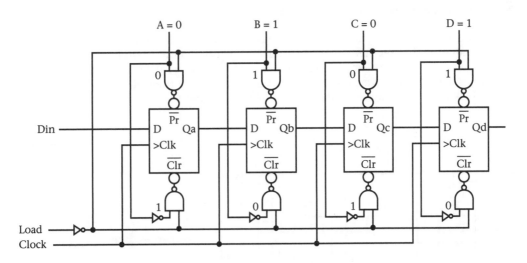

(A) Place the number 10 at the inputs DCBA (Note: 10 = 1010).

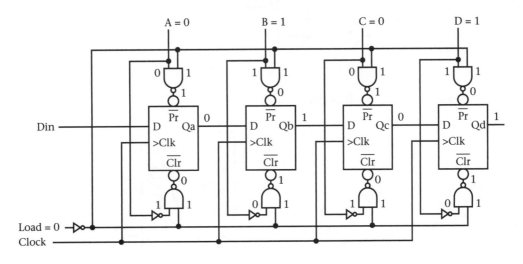

(B) Connect LOAD to 0.

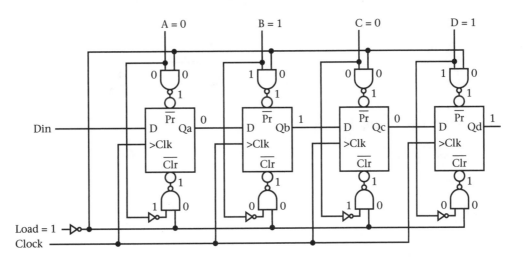

(C) Connect LOAD to 1.

Figure 4-2 4-Bit Loadable Shift Register

Recirculating a Shift Register

Figure 4-3 shows the symbol for a loadable 4-bit shift register and how to use it to build a 4-bit recirculating shift register. A recirculating shift register connects Qd back to Din (the input of Qa). The binary data is shifted in closed loop or ring. The binary data completes the loop after 4 clock pulses. Figure 4-4 shows how the number 9 is loaded into the shift register. Figure 4-5 shows how 4 clock pulses recirculate the number 9 one time around the shift register.

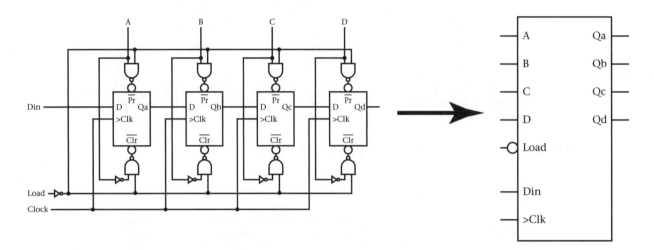

(A) Symbol for a 4-bit loadable shift register.

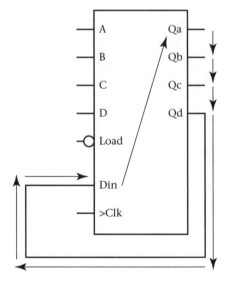

(B) Building a 4-bit recirculating shift register.

Figure 4-3 Shift Register Symbols

Step 1: Initial conditions

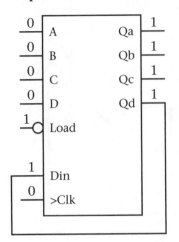

Step 2: Place the number 9 at inputs ABCD.

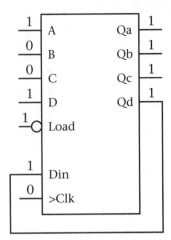

Step 3: Connect Load to 0. This loads the number 9 to outputs Qa, Qb, Qc, Qd.

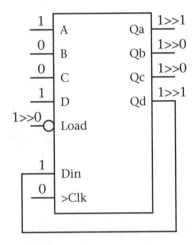

Step 4: Connect Load to 1. Din and Clk can now be used to control the shift register. ABCD no longer have any effect.

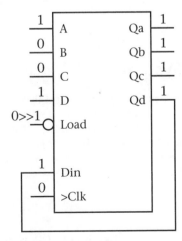

Figure 4-4 Loading the number 9 into the shift register

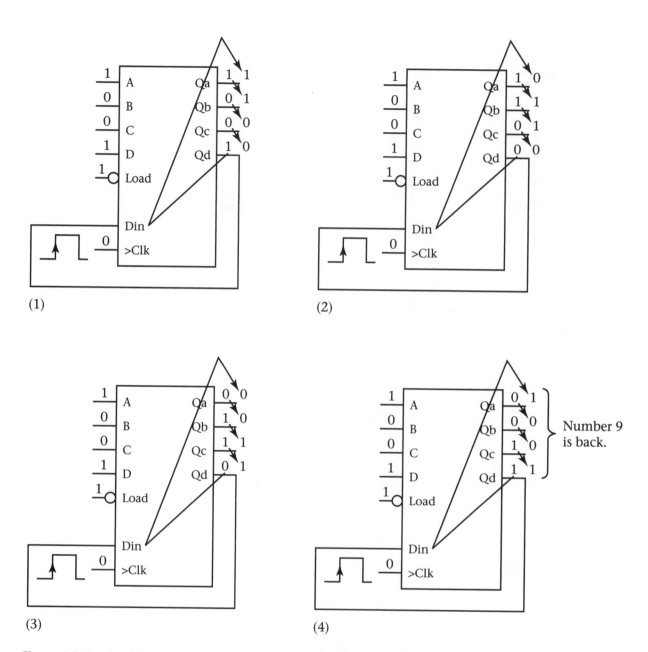

Figure 4-5 Use the shift register to recirculate the number 9.

Serial Data Transfer System

Digital systems communicate using a series of 1s and 0s. Two shift registers can be used to transfer data from one digital system to another as a series of bits. Refer to Figure 4-6. The number to be transferred is loaded into the transmitter shift register. The connection from Qd to Din is called a serial link. The link is used to transfer the 4-bit number to the receiver. The serial transfer requires 4 clock pulses to shift the 4-bit number. A serial transfer system allows a large number of bits to be sent over a small number of wire connections.

Latch Fundamentals

A **Latch** is similar to a **D** flip-flop. The clock input >**Clk** is replaced with **Ena** (Enable). Unlike D flip-flops, the latch is continuously asserted when **Ena.** is connected to logic level 1. D is

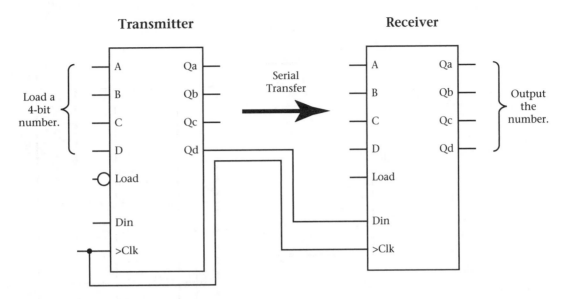

Figure 4-6 Serial Data Transfer System

continuously transferred to Q while Ena. = 1. A latch responds to levels at "Ena." and not clock
edges. The > symbol represents edge triggering and is removed from the "Ena." input. Refer to
Figure 4-7.

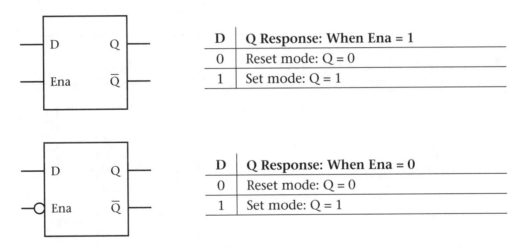

D	Q Response: When Ena = 1
0	Reset mode: Q = 0
1	Set mode: Q = 1

D	Q Response: When Ena = 0
0	Reset mode: Q = 0
1	Set mode: Q = 1

Figure 4-7 Active High Latch and Active Low Latch

Data Registers

Register means "record." A data register is a device that can be used to record binary data.
Recording binary data is also synonymous with storing binary data (memory). Latches and D flip-
flops can each be used to construct two different types of data registers. Figure 4-8 shows an
example of a 4-bit data register made up from four latches and a second data register made up of
four D flip-flops.

With the latch data register, the 4-bit number to be registered into memory is placed at
inputs D3, D2, D1, D0. The enable input is asserted with a 1 and the number is stored at outputs
Q3, Q2, Q1, Q0. When the enable input is connected to 0, the D inputs cannot change the Q
outputs. The latches are no longer enabled.

With the D flip-flop data register, the 4-bit number to be registered into memory is placed at
inputs D3, D2, D1, D0. The clock input is asserted with a positive edge (0 to 1 transition) and the

Data Register Using Latches

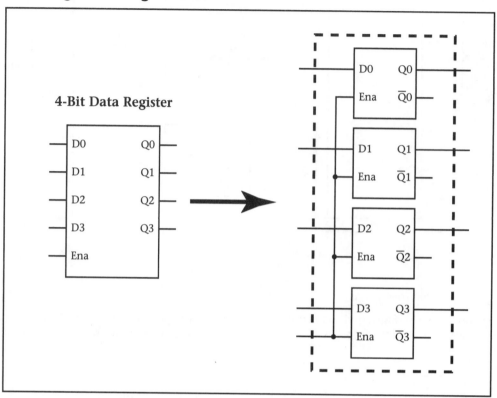

Data Register Using D Flip-Flops

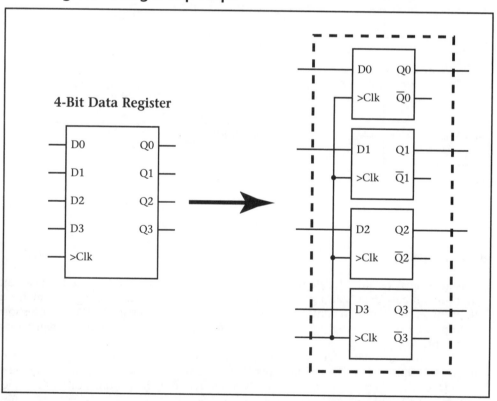

Figure 4-8 4-Bit Data Register

number is stored at outputs Q3, Q2, Q1, Q0. A positive edge is required to transfer or store a number. A negative edge, a constant "1," or a constant "0" at the clock input will not change the data at the "Q" outputs.

Parallel Data Transfer System

Two data registers can be used to transfer data from one digital system to another. Because all 4 bits are transferred at the same time it is called parallel transfer. Refer to Figure 4-9.

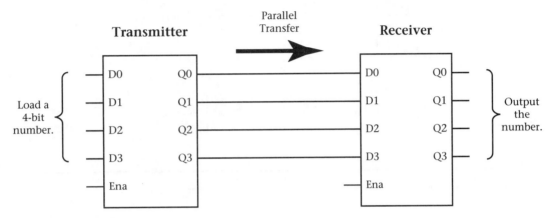

Figure 4-9 Parallel Data Transfer System

The number to be transferred is loaded into the transmitter data register. The four connections from Q3, Q2, Q1, Q0 to D3, D2, D1, D0 are called a parallel link. The link is used to transfer the 4-bit number to the receiver. Parallel transfer allows all bits to be sent at the same time. Each bit requires a wire connection.

POWERPOINT PRESENTATION

Use PowerPoint to view the Lab 04 slide presentation.

LAB WORK PROCEDURE

Design a Serial Data Transfer System

You will use two UP boards to test a serial data transfer system. The first UP board will be the transmitter and the second will be the receiver. A pushbutton switch connected to the clock inputs of the transmitter and receiver will control the data transfer. The pushbutton will need to be debounced. Linking the two UP boards requires wiring access to the Altera IC. UP boards are shipped from the factory with wire headers on the MAX IC but not on the FLEX IC. The FLEX IC requires that you solder a flex expansion header socket to the UP board. Appendix C describes the FLEX expansion header in more detail.

Transmitter for the Serial Data Transfer System

1. You will draw a diagram for a transmitter. Use the Lab Work Procedure in Lab 1 as a guide and complete steps 1 through 12. Save all files in a folder (directory) named **Labs**. Use the project name **lab4t**. The D flip-flop symbol in the "prim" library is **DFF**. The NAND gate symbol is "**NAND2**." The debounce symbol will appear in the symbol library because of the work you completed in the previous lab. The Graphic Editor file for the transmitter should resemble Figure 4-10.

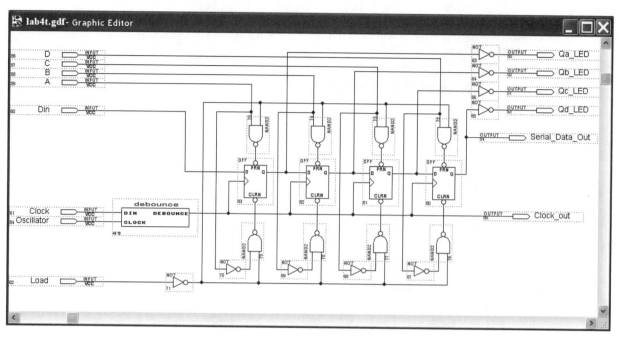

Figure 4-10 Altera "Graphic Editor": Transmitter for the Serial Data Transfer System

> *NOTE: Rotating a symbol: right click on a symbol and the menu will allow you to rotate it.*

> *NOTE: Drawing lines too close to a symbol: lines crossing at the edge of a symbol boundary box (the box drawn with green dotted lines) will be connected with a connection dot. Be sure to space out your symbols and not run lines next to or through a symbols boundary box.*

2. Lab 1 describes all the steps that are required to program a UP board with the transmitter. Use the Lab Work Procedure in Lab 1 as a guide and complete steps 13 through 17. Ignore any compiler warning and information messages.

MAX and FLEX IC Considerations

MAX IC Design:
- Connect input **Load**, **A**, **B**, **C**, and **D** to a UP board MAX switches.
- Connect input **Clock** to a UP board MAX pushbutton.
- Assign input **Oscillator** to pin **83**. A PCB trace connects the 25,175,000 PPS oscillator to pin 83 of the MAX IC. Refer to Lab 1, step 13_{FLEX} for assistance in assigning an input symbol to a pin number.
- Connect outputs **Qa_LED**, **Qb_LED**, **Qc_LED**, and **Qd_LED** to UP board MAX LEDs.
- *Do not* connect Din, Serial_data_out, and Clock_out at this time. The connections are described in the Lab Exercise section.

FLEX IC Design:
- Assign input **Load**, **A**, **B**, **C**, and **D** to a UP board FLEX switch.
- Assign input **Clock** to pin **28** or pin **29** to connect to a FLEX pushbutton.
- Assign input **Oscillator** to pin **91**. A PCB trace connects the 25,175,000 PPS oscillator to pin 91 of the FLEX IC.
- Assign outputs **Qa_LED**, **Qb_LED**, **Qc_LED**, and **Qd_LED** to UP board FLEX LEDs. Remember to turn off unused LEDs by connecting them to Vcc.
- Assign input **Din** and outputs **Serial_data_out** and **Clock_out** to pins on the FLEX expansion header. Refer to Appendix C. Do not connect wires to these pins at this time. The connections are described in the Lab Exercise section.

3. At this point you have drawn the diagram, programmed the IC, and connected all switches and LEDs. The transmitter is ready for testing.

Receiver for the Serial Data Transfer System

1. You will draw a diagram for a receiver. Use the Lab Work Procedure in Lab 1 as a guide and complete steps 1 through 12. Save all files in a folder (directory) named **Labs**. Use the project name **lab4r**. The D flip-flop symbol in the "prim" library is **DFF**. The Graphic Editor file for the receiver should resemble Figure 4-11.

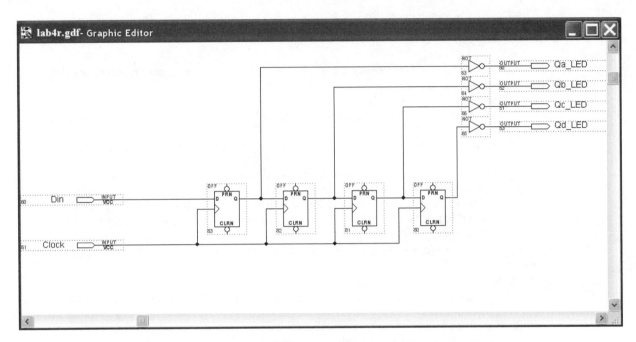

Figure 4-11 Altera "Graphic Editor": Receiver for the Serial Data Transfer System

2. Lab 1 describes all the steps that are required to program a UP board with the receiver. Use the Lab Work Procedure in Lab 1 as a guide and complete steps 13 through 17. Ignore any compiler warning and information messages.

MAX and FLEX IC Considerations

MAX IC Design:
- Connect **Qa_LED**, **Qb_LED**, **Qc_LED**, and **Qd_LED** to UP board LEDs.
- The UP board oscillator is connected to pin 83 of the MAX IC. This pin number may be assigned to clock of the DFF. If this happens, then the DFF will receive 25,175,000 clock pulses per second. To make sure this does not happen, assign pin 2 to the clock input. Refer to Lab, 1 step 13$_{FLEX}$ for assistance regarding assigning an input symbol to a pin number.
- *Do not* connect "Din" and "Clock." The connections are described in the Lab Exercise section.

FLEX IC Design
- Assign outputs **Qa_LED**, **Qb_LED**, **Qc_LED**, and **Qd_LED** to UP board FLEX LEDs. Remember to turn off unused LEDs by connecting them to Vcc.
- Assign **Din** and **Clock** to pins on the FLEX expansion header. Refer to Appendix C.
- *Do not* connect "Din" and "Clock." The connections are described in the Lab Exercise section.

3. At this point you have drawn the diagram, programmed the IC, and connected all LEDs. The receiver is ready for testing.

LAB EXERCISES AND QUESTIONS

This section contains lab exercises that can be performed on the UP board and questions that can be answered at home. Ask your instructor which exercises to perform and which questions to answer.

Lab Exercises

Exercise 1: Transmitter test #1: Use only the transmitter system to test basic shift register operation.
Try the following test on the transmitter board:
1. Connect Din to a switch.
2. Disable the Load input and use clock and Din to shift and to fill the transmitter with 1s.
3. Disable the Load input and use clock and Din to shift and to fill the transmitter with 0s.
4. Use DCBA and Load switches to load the number 6 into the transmitter.
   a. In step 2, how many clock pulses are required to fill the transmitter with 1s?
   b. How many clock pulses are required to load the number 6 into the transmitter?
   c. What happens if you leave DCBA = 6, the load input low, and you try to fill the transmitter with 1s?

Exercise 2: Transmitter test #2: Use only the transmitter system to test a recirculating shift register system.
Try the following test on the transmitter board:
1. Connect a wire from **Serial_Data_Out** to **Din**. This will create a recirculating shift register system.
2. Use DCBA and Load switches to load the number 9 into the transmitter.
3. Use the clock input to recirculate the number 9. Generate and record your observations in a table similar to the one shown in Figure 4-12. Once the table is filled, disconnect the wire from Serial_Data_Out to Din.

Step #	Load	D C B A	Din	Clock	Qd Qc Qb Qa	Description
1					1 0 0 1	Load #9
2						Recirculate clock pulse 1
3						Recirculate clock pulse 2
4						Recirculate clock pulse 3
5					1 0 0 1	Recirculate clock pulse 4

Figure 4-12 Test Result Table for a Recirculating Shift Register

Exercise 3: Test a serial data transfer system.
You will link the transmitter to the receiver and test the operation of a serial data transfer system.
   Try the following test on the serial data transfer system:
1. Connect three long wires from the transmitter to the receiver.
   a. Connect a wire from **Serial_Data_Out** of the transmitter to **Din** of the receiver.
   b. Connect a wire from the **Clock_Out** output of the transmitter to the **Clock** of the receiver.
   c. Connect a wire to ground the transmitter and the receiver.
   The pin in the lower right-hand corner of the JTAG_OUT socket is ground. Connect a long wire from the **JTAG_OUT** socket of the transmitter to **JTAG_OUT** socket of the receiver. This ground wire is required whenever logic levels are shared between two different systems with separate power packs. Refer to Figure 4-13.
2. Use the serial data transfer system to transfer the number 9 from the transmitter board to the receiver board. Generate and record your observations in a table similar to the one shown in Figure 4-14.

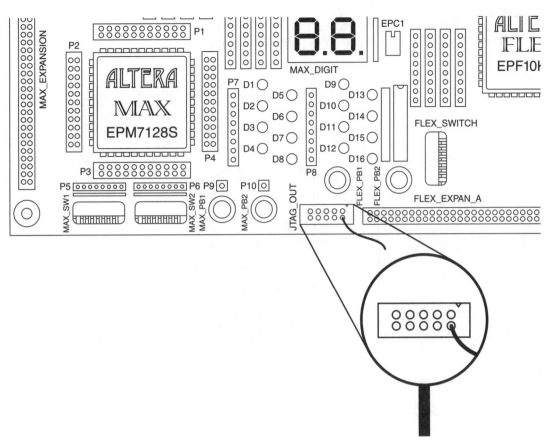

Figure 4-13 Use the JTAG connecter to ground the transmitter and the receiver.

Step #	D C B A	Load	Clock	Transmitter Qd Qc Qb Qa	Receiver Qd Qc Qb Qa
1					
2					
3					
4					
5					

Figure 4-14 Test Result Table for the Serial Data Transfer System

> *NOTE: Long wires can cause excessive noise and timing problems. If the system does not function properly, connect a single NOT gate to the clock input of the receiver UP board. This will make the receiver UP board receive data on the negative edge of the clock while the transmitter UP board continues to transmit data on the positive edge of the clock. This will allow the system to be more stable because only one UP board is clocked at a time.*

3. Use the UP boards to transfer a number *other than 9* from the transmitter board to the receiver board. Extend the table and record your observations.
4. Make the transmitter into a recirculating shift register and try another transfer. Extend the table shown in Figure 4-14 and record your observations.

Exercise 4: Design a parallel data transfer system.

Change the **serial data transfer system** to a **parallel data transfer system**. A parallel transfer system uses a 4-bit data register. The transmitter and receiver systems are identical. A 4-bit data register is made up of four latches. The latch symbol in the "prim" library is **latch**. The Graphic Editor file for the 4-bit data register system should resemble Figure 4-15.

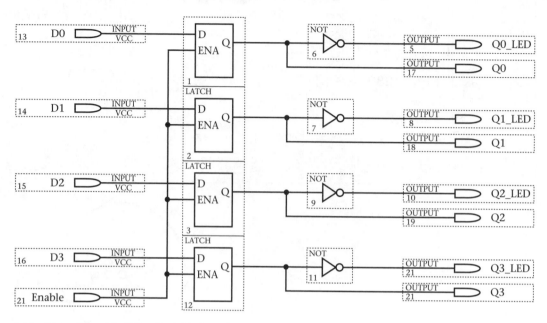

Figure 4-15 Altera "Graphic Editor": 4-Bit Data Register System

Try the following test on the parallel data transfer system:
1. Connect five long wires from the transmitter to the receiver.
   a. Connect four wires from Q3, Q2, Q1, Q0 of the transmitter to D3, D2, D1, D0 of the receiver.
   b. Connect a wire from the ground pin of the transmitter to the ground pin of the receiver. Refer to Figure 4-13 shown in Exercise 3.
2. Use the UP boards to transfer the number 9 from the transmitter board to the receiver board. Generate and record your observations in a table similar to the one shown in Figure 4-16.

Step #	Transmitter D3 D2 D1 D0	Tx Ena	Transmitter Q3 Q2 Q1 Q0	Rx Ena	Receiver Q3 Q2 Q1 Q0
1					
2					

Figure 4-16 Test Result Table for the Parallel Data Transfer System

3. Use the UP boards to transfer a number *other than 9* from the transmitter board to the receiver board. Extend the table shown in Figure 4-16 and record your observations.
4. Why is switch bounce not a problem for the parallel transfer system?

Lab Questions

1. Fill in the table to show how to load the number 6 into the shift register and recirculate the number one time around the shift register. Refer to Figure 4-17.

4-Bit Shift Register

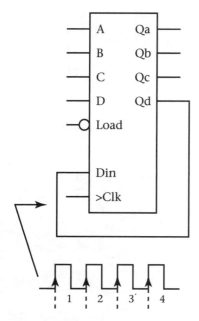

Clock Pulse	Load	D	C	B	A	Din	Qd	Qc	Qb	Qa
Load #6 ->										
Before +ve edge 1										
After +ve edge 1										
Before +ve edge 2										
After +ve edge 2										
Before +ve edge 3										
After +ve edge 3										
Before +ve edge 4										
After +ve edge 4										

Figure 4-17 Shift Register System for Question 1

2. Connect the transmitter to the receiver to create a serial data transfer system. Make the transmitter a recirculating shift register. Fill in the table to show how to load the number 9 into the transmitter and transfer the number to the receiver. Refer to Figure 4-18.

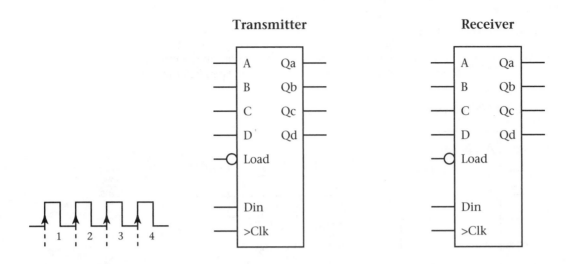

Figure 4-18 Serial Data Transfer System for Question 2

Clock Pulse	Transmitter Shift Register										Receiver Shift Register									
	Load	D	C	B	A	Din	Qd	Qc	Qb	Qa	Load	D	C	B	A	Din	Qd	Qc	Qb	Qa
Load #9 into trans											1	X	X	X	X		0	0	0	0
Before Clk edge 1																				
After Clk edge 1																				
Before Clk edge 2																				
After Clk edge 2																				
Before Clk edge 3																				
After Clk edge 3																				
Before Clk edge 4																				
After Clk edge 4																				

3. Connect U1 to U2 to create a bidirectional serial data transfer system. Bidirectional means each shift register can be used as a transmitter or a receiver. Data will transfer simultaneously from U1 to U2 and from U2 to U1. Fill in the table to show how the system would work if the number 7 is loaded into U1 and the number 5 into U2. Refer to Figure 4-19.

4. An 8-bit serial data transfer system requires two 8-bit shift registers. If an 8-bit transmitter and an 8-bit receiver were constructed on UP boards, how many long wires would be required to connect the two UP boards? How many clock pulses are required to transfer an 8-bit number? The answer to the question highlights the operational characteristics of serial data transfer systems.

5. An 8-bit parallel transfer system requires two 8-bit data registers. If an 8-bit transmitter and an 8-bit receiver were constructed on UP boards, how many long wires would be required to connect the two UP boards? How many clock pulses are required to transfer an 8-bit number? The answer to the question highlights the operational characteristics of parallel data transfer systems.

6. Assume a 64-bit transfer system is required. The designer must choose serial or parallel transfer.
   a. Which transfer system would likely be fastest at transferring a 64-bit number? Serial or parallel?
   b. Which 64-bit transfer system requires the fewest wire connections between the transmitter and receiver? Serial or parallel?

7. A computer keyboard uses a serial data transfer system to send an 8-bit data code to the PC when a user presses a key. What is the advantage of using a serial data transfer system in this application?

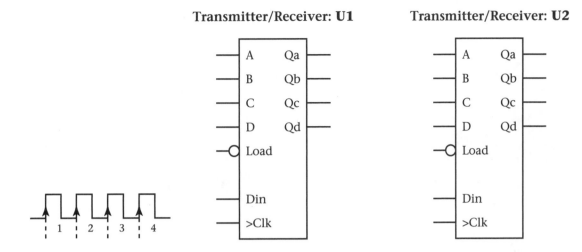

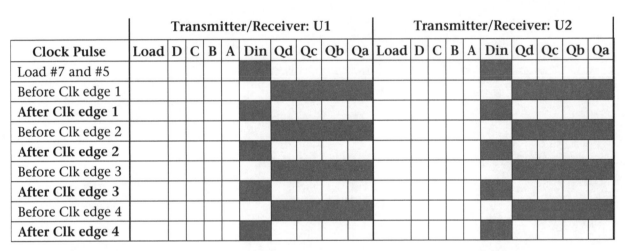

Clock Pulse	Transmitter/Receiver: U1										Transmitter/Receiver: U2									
	Load	D	C	B	A	Din	Qd	Qc	Qb	Qa	Load	D	C	B	A	Din	Qd	Qc	Qb	Qa
Load #7 and #5																				
Before Clk edge 1																				
After Clk edge 1																				
Before Clk edge 2																				
After Clk edge 2																				
Before Clk edge 3																				
After Clk edge 3																				
Before Clk edge 4																				
After Clk edge 4																				

Figure 4-19 Serial Data Transfer System for Question 3

Conclusions

- $\overline{Pr}$ and $\overline{Clr}$ are asynchronous inputs that are used to set initial conditions at Q.
- A loadable shift register uses Pr and Clr to load a number.
- A 4-bit recirculating shift register requires 4 clock pulses to shift a number in a complete loop.
- A serial data transfer system is useful in transferring large quantities of bits over long distances because it uses few wires.
- A latch continually transfers D to Q when the enable input is asserted.
- Parallel data transfer is fast but it does require a wire for each bit.

Lab 5

JK Flip Flop and Counter Fundamentals

Equipment

Altera UP board
Book CD-ROM
Spool of 24 AWG wire
Wire cutters
Blank disk

Objectives

Upon completion of this lab, you should be able to:

- Operate a JK flip-flop.
- Operate a 4-bit counter system with load, cascade, count up, and count down features.
- Cascade counter symbols.
- Design and operate a frequency divider system.
- Design and operate a VHDL binary counter.
- Design and operate a VHDL BCD counter.

INTRODUCTORY INFORMATION

In Lab 4, D flip-flops were used as data registers and shift registers to store and transfer binary data. In this lab you will learn how to use a new device called a JK flip-flop to create a digital system capable of counting in binary.

JK Flip-Flop Fundamentals

The JK flip-flop is an improved SR flip-flop. The Ambiguous mode is replaced by the Toggle mode. The Set, Reset, and No Change modes are still used. Refer to Figure 5-1.

113

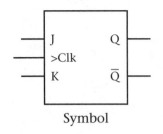

J	K	Q Response: On +ve Edge of Clock
0	0	No change mode. Hold mode.
0	1	Reset mode: Q = 0
1	0	Set mode: Q = 1
1	1	Toggle mode. Invert output Q.

Symbol

Figure 5-1 Clocked JK Flip-Flop

To help remember the **RESET** mode, you can think of **K = 1** means Kill the output.

To help remember the **SET** mode, you can think of **J = 1** means the output Jumps up to 1.

The **TOGGLE** mode changes the output logic level. If Q started at 0, then Q changes to 1 after the clock is asserted. If Q started at 1, then Q changes to 0 after the clock is asserted. Figure 5-2 shows the inside of the JK flip-flop.

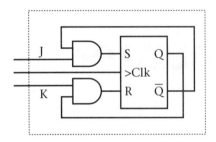

J = K = 1 will set the flip-flop if the initial condition at Q = 0.
J = K = 1 will reset the flip-flop if the initial condition at Q = 1.
A JK flip-flop in the toggle mode is like a power button on a car radio. You press the button once to turn the radio on. You press the same button again to turn the radio off. You toggle the radio on and off.

Construction Diagram

Figure 5-2 JK Flip-Flop Construction Diagram

Counter Fundamentals

The counters used with Altera ICs are called parallel or synchronous counters. The counter theory, presented next, is called ripple or asynchronous counter theory. Ripple counter theory is functionally identical to parallel counter theory but it is easier to understand. Ripple counters are better suited to introduce counters to students. Parallel counters will be presented later.

Negative edged triggered JK flip-flops in the toggle mode can be connected together to create a binary counter system. Figure 5-3 demonstrates the operation of three JK flip-flops connected together.

Figure 5-3 shows that a clock pulse is applied to the clock input of Qa. Qa toggles at each negative edge transition of the clock input. The signal generated at Qa becomes the clock signal for Qb. Qb in turn toggles for each negative edge transition of Qa. The signal generated at Qb becomes the clock signal for Qc. Qc in turn toggles for each negative edge transition of Qb. This chain reaction can be viewed as a ripple effect. The clock signal ripples from Qa through to Qc.

A counter system counts clock pulses. Let's assume the clock rate is 1 PPS. This means there is a 1-second time interval between each negative edge of the clock. Analyzing each 1-second time interval and placing the response at Qa, Qb, Qc in a table will reveal the binary count sequence at Qa, Qb, Qc. Figure 5-4 shows the analysis required to fill in the binary count table.

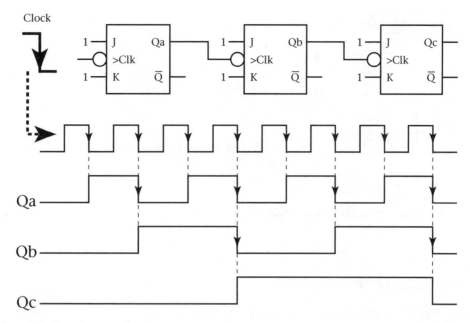

Figure 5-3 Three-Flip-Flop Counter System

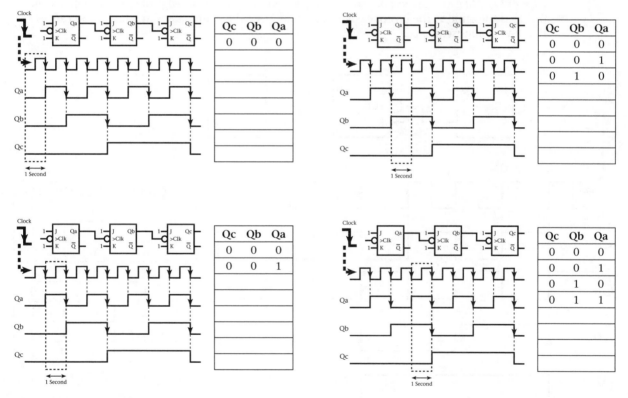

Figure 5-4 Count Table for a Three-Flip-Flop Counter System

Continue the waveform analysis to complete the count table. Refer to Figure 5-5.

The input clock controls the speed at which the counter counts. A 1 PPS input clock will generate a count sequence from 0 to 7 in 8 seconds. Each count state lasts 1 second. If the clock were set to 1,000 PPS, then a count cycle would only take 8 milliseconds. If LEDs were to be connected to the counter, it would result in the appearance that all LEDs are on at the same time. At this PPS rate, the count sequence is too fast to be visible on the three LEDs. We are only capable of seeing an LED blink if the pulse rate is less than 30 PPS.

Count	Qc	Qb	Qa
0	0	0	0
1	0	0	1
2	0	1	0
3	0	1	1
4	1	0	0
5	1	0	1
6	1	1	0
7	1	1	1

The table is called a COUNT state table.

The counter is called a MOD 8 counter because it has 8 different count states.

MOD is short for the word MODULUS.

After 7 the counter restarts from 0 and the cycle is repeated.

Figure 5-5 Complete Count Table for a Three-Flip-Flop Counter System

Down Counters

Up counters count forward and down counters count backwards. To make a counter count backwards, all you need to do is connect the $\overline{Q}$ to the Clk of the next flip-flop. Refer to Figure 5-6. Qa toggles on every negative edge of the clock input. Qb toggles on every negative edge of $\overline{Q}a$ ($\overline{Q}a$ is the inverse of Qa). Qc toggles on every negative edge of $\overline{Q}b$. A negative edge on $\overline{Q}b$ is the same as the positive edge Qb. If you place the count states in a table, you can see the down count sequence.

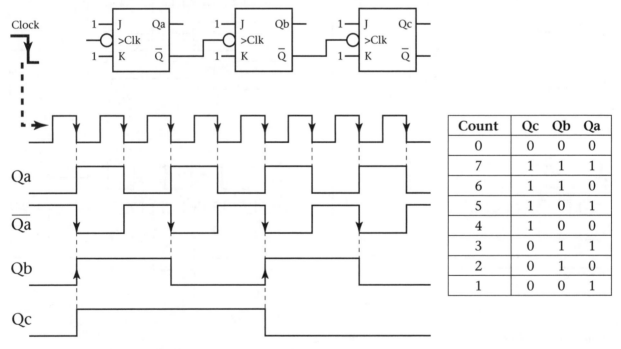

Count	Qc	Qb	Qa
0	0	0	0
7	1	1	1
6	1	1	0
5	1	0	1
4	1	0	0
3	0	1	1
2	0	1	0
1	0	0	1

Figure 5-6 Down Counter System

Asynchronous and Synchronous Presettable Counters

Presettable means the counter can be loaded with a number that represents the starting point of the count sequence. *Asynchronous* means the counter is loaded without an asserted clock. In Lab 4 you learned how the asynchronous inputs were used to load a number into a shift register. The theory described in Lab 4 also applies to the asynchronous loading of a counter. *Synchronous* means the clock must be asserted to load a number into the counter. The diagram in Figure 5-7 shows how the clock must be asserted to complete the loading cycle.

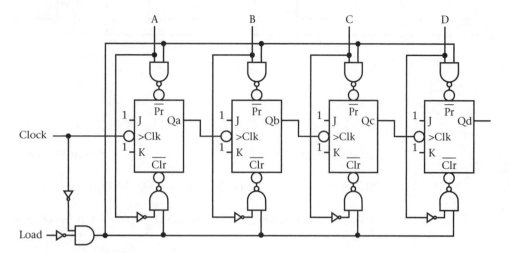

Figure 5-7 Synchronous Presettable Counter

Combination Up/Down Counter

The system shown in Figure 5-8 combines the features of both an up counter and a down counter. The system has a count direction control input to select up counting or down counting. When the control input is 0, the top AND gates will pass the logic levels from the Q outputs to the clock of the next flip-flop. A "Q" to clock connection creates an up counter! When the control input is 1, the bottom AND gates will pass the logic levels from the $\overline{Q}$ outputs to the clock of the next flip-flop. A $\overline{Q}$ to clock connection creates a down counter!

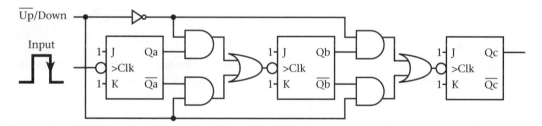

Figure 5-8 Combination Up/Down Counter System

Parallel Counters

Ripple counters use JK flip-flops in the toggle mode. Each flip-flop's clock is chained to the Q output of the previous flip-flop stage. The architecture used inside an Altera IC is designed to work with counter systems that clock all flip-flop stages at the same time. Parallel counters are built from JK flip-flops that are all clocked at the same time. Refer to Figure 5-9.

Figure 5-9 describes the operation of the parallel counter. Parallel counters are a little more complex to construct because they need additional AND gates.

Cascading Counters

Figure 5-10 is a parallel counter that has an input **Cin** and an output **Cout**. Cin = 1 allows the counter to count. Cin = 0 holds the counter from counting. Cout = 1 when **terminal count** 15 is

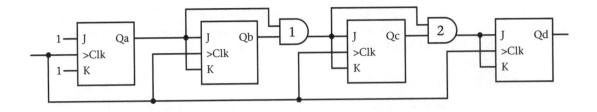

Count Table

Qd	Qc	Qb	Qa
0	0	0	0
0	0	0	1
0	0	1	0
0	0	1	1
0	1	0	0
0	1	0	1
0	1	1	0
0	1	1	1
1	0	0	0
1	0	0	1
1	0	1	0
1	0	1	1
1	1	0	0
1	1	0	1
1	1	1	0
1	1	1	1

Study column Qa. Qa toggles on every clock pulse.

Study column Qb. Qb only toggles if Qa = 1. Qb holds when Qa = 0.

Study column Qc. Qc only toggles if Qa **AND** Qb = 1.

Study column Qd. Qd only toggles if Qa **AND** Qb **AND** Qc = 1.

AND gate #1

AND gate #2

Figure 5-9 Parallel Counter

reached. *Terminal count* is the last count that is generated by the counter before it recycles back to its starting point. Cout = 0 when the count is in the range of 0 to 14.

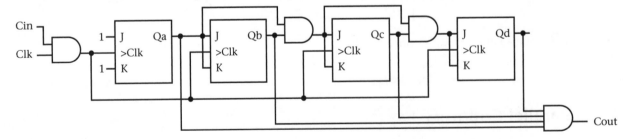

Figure 5-10 Parallel Counter with Cin and Cout

Cin and Cout allow counters to be cascaded. *Cascade* means "chain counter stages to each other." For example, to build an 8-bit counter you would need to cascade two 4-bit counters. **Cin** (Carry **in**put) would be connected to **Cout** (Carry **out**put). Refer to Figure 5-11.

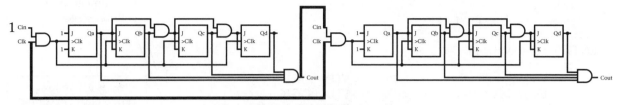

Figure 5-11 Two 4-bit counters cascaded to make an 8-bit counter

The second 4-bit stage counts forward when Cout of the first stage is 1. This occurs when the first stage reaches terminal count 15. The 8-bit counter is Mod 256. It counts from 0 to 255 in binary. The Mod number is calculated by multiplying the Mod number of each stage ($16 \times 16 =$ Mod 256).

VHDL Binary Counter

Figure 5-12 shows the code for a 4-bit VHDL binary counter. The comments surrounding the VHDL code explain its operation.

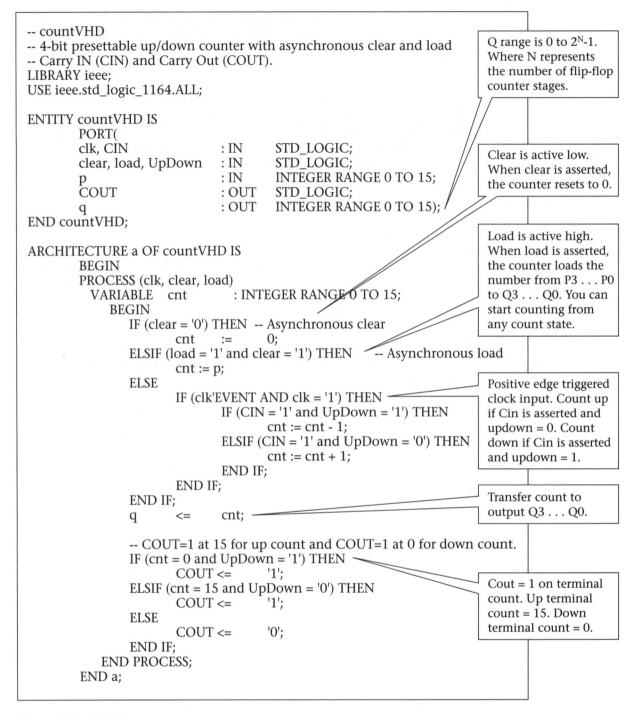

```
-- countVHD
-- 4-bit presettable up/down counter with asynchronous clear and load
-- Carry IN (CIN) and Carry Out (COUT).
LIBRARY ieee;
USE ieee.std_logic_1164.ALL;

ENTITY countVHD IS
        PORT(
        clk, CIN                : IN        STD_LOGIC;
        clear, load, UpDown     : IN        STD_LOGIC;
        p                       : IN        INTEGER RANGE 0 TO 15;
        COUT                    : OUT       STD_LOGIC;
        q                       : OUT       INTEGER RANGE 0 TO 15);
END countVHD;

ARCHITECTURE a OF countVHD IS
        BEGIN
        PROCESS (clk, clear, load)
          VARIABLE    cnt           : INTEGER RANGE 0 TO 15;
            BEGIN
                IF (clear = '0') THEN  -- Asynchronous clear
                        cnt    :=      0;
                ELSIF (load = '1' and clear = '1') THEN     -- Asynchronous load
                        cnt := p;
                ELSE
                        IF (clk'EVENT AND clk = '1') THEN
                                IF (CIN = '1' and UpDown = '1') THEN
                                        cnt := cnt - 1;
                                ELSIF (CIN = '1' and UpDown = '0') THEN
                                        cnt := cnt + 1;
                                END IF;
                        END IF;
                END IF;
                q       <=      cnt;

                -- COUT=1 at 15 for up count and COUT=1 at 0 for down count.
                IF (cnt = 0 and UpDown = '1') THEN
                        COUT <=       '1';
                ELSIF (cnt = 15 and UpDown = '0') THEN
                        COUT <=       '1';
                ELSE
                        COUT <=       '0';
                END IF;
            END PROCESS;
        END a;
```

Callout boxes:

- Q range is 0 to 2^N-1. Where N represents the number of flip-flop counter stages.

- Clear is active low. When clear is asserted, the counter resets to 0.

- Load is active high. When load is asserted, the counter loads the number from P3 . . . P0 to Q3 . . . Q0. You can start counting from any count state.

- Positive edge triggered clock input. Count up if Cin is asserted and updown = 0. Count down if Cin is asserted and updown = 1.

- Transfer count to output Q3 . . . Q0.

- Cout = 1 on terminal count. Up terminal count = 15. Down terminal count = 0.

Figure 5-12 VHDL Binary Counter

VHDL BCD Counter

A **BCD** counter is a counter that counts in a code that is equivalent to decimal. BCD is an acronym for the words Binary Coded Decimal. A BCD number is a 4-bit number that is used to represent the decimal numbers 0 to 9. Refer to Figure 5-13.

To convert a decimal number to BCD you must not use the binary PWC. You simply convert each decimal numeral to a 4-bit BCD number. For example, to convert 25 to BCD you combine the 4-bit BCD code for 2 with the 4-bit BCD code for 5. The result is 0010 0101. Compare this answer to the answer for converting the number 25 to binary in Lab 1 and you will see the difference.

Decimal	BCD			
0	0	0	0	0
1	0	0	0	1
2	0	0	1	0
3	0	0	1	1
4	0	1	0	0
5	0	1	0	1
6	0	1	1	0
7	0	1	1	1
8	1	0	0	0
9	1	0	0	1

The 4-bit numbers above 9 are not BCD numbers.

Decimal	4-bit Binary			
10	1	0	1	0
11	1	0	1	1
12	1	1	0	0
13	1	1	0	1
14	1	1	1	0
15	1	1	1	1

Figure 5-13 BCD Numbers

A counter that counts from 0 to 9 is a MOD 10 counter or a BCD counter. To build a MOD 10 counter you begin with a binary counter that includes the range of count states 0 to 9. A MOD 16 counter includes the range from 0 to 9 and uses four flip-flop stages. You then use counter feedback or state diagram techniques to alter the binary counter. Counter feedback and state diagram techniques have been used with older generation IC technology for many years to modify binary counters. These techniques often require a large number of logic gates to be connected to the binary counter in order to alter the count sequence. The resulting counter system can often be complex to draw and difficult to analyze and implement. A more modern solution is to apply the same design techniques to alter the code of a VHDL binary counter. The code modification technique uses VHDL "IF" statements as feedback to alter the natural count sequence progression of a VHDL binary counter. The results are the same but the complexity of the design is reduced.

Figure 5-14 shows the code for a VHDL BCD counter. The comments surrounding the VHDL code explain its operation.

```
-- countBCD
-- 4-bit presettable up/down BCD counter with asynchronous clear and load
-- and Carry IN (CIN) and Carry Out (COUT).
LIBRARY ieee;
USE ieee.std_logic_1164.ALL;

ENTITY countBCD IS
        PORT(
        clk, CIN                : IN    STD_LOGIC;
        clear, load, UpDown     : IN    STD_LOGIC;
        p                       : IN    INTEGER RANGE 0 TO 15;
        COUT                    : OUT   STD_LOGIC;
        q                       : OUT   INTEGER RANGE 0 TO 15);
END countBCD;

ARCHITECTURE a OF countBCD IS
        BEGIN
        PROCESS (clk, clear, load)
        VARIABLE   cnt     : INTEGER RANGE 0 TO 15;
           BEGIN
                IF (clear = '0' ) THEN     -- Asynchronous clear
                        cnt      :=     0;
                ELSIF (load = '1' and clear = '1') THEN-- Asynchronous presettable load
                        cnt := p;

                ELSE
                IF (clk'EVENT AND clk = '1') THEN
                        IF (CIN = '1' and UpDown = '1') THEN
                                cnt := cnt - 1;
                                if (cnt =15) then cnt := 9; END IF;
                        ELSIF (CIN = '1' and UpDown = '0') THEN
                                cnt := cnt + 1;
                                if (cnt =10) then cnt := 0; END IF;
                                END IF;
                        END IF;
                END IF;
                q       <=      cnt;

                -- COUT=1 at 9 for up count and COUT=1 at 0 for down count.
                IF (cnt = 0 and UpDown = '1') THEN
                        COUT <=         '1';
                ELSIF (cnt = 9 and UpDown = '0') THEN
                        COUT <=         '1';
                ELSE
                        COUT <=         '0';
                END IF;
           END PROCESS;
        END a;
```

> Q range is 0 to 2^N-1. Where N represents the number of flip-flop counter stages. It must be a range that defines a binary counter.

> When counting backwards 3, 2, 1, and 0, the next count in the binary sequence is 15. The "IF" statement causes the counter to count backwards from 9.

> When counting up 6, 7, 8, and 9, the next count in the binary sequence is 10. The "IF" statement causes the counter to count forward from 0.

> When counting up 6, 7, 8, and 9, the terminal count is 9.

Figure 5-14 VHDL BCD Counter

Frequency Division

Frequency of a pulse waveform is the pulse rate. The oscillator on the UP board generates a pulse waveform at a rate of 25,175,000 PPS. This is the frequency or the pulse rate of the oscillator. The frequency of the oscillator can be slowed down (or divided down) by a counter system.

Figure 5-15 shows how a MOD 8 counter can be used as a frequency divider system.

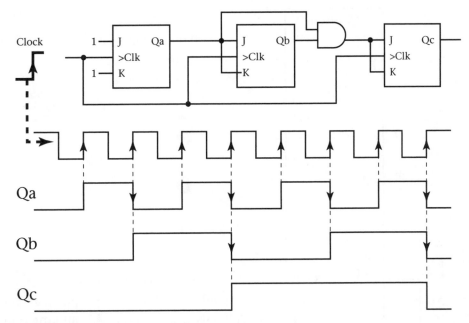

Figure 5-15 MOD 8 Counter as a Frequency Divider System

The pulse rate at output Qa is half the pulse rate of the clock. The pulse rate at output Qb is half the pulse rate of the Qa. The pulse rate at output Qc is half the pulse rate of the Qb. If the pulse rate at the clock is 8,000 PPS, then the pulse rate at Qa is 4,000 PPS. In general a MOD 8 counter will divide the frequency of the clock by 8. Qc pulse rate is 8,000/8 = 1,000 PPS. A MOD 8 counter can also be called a "divide by 8 counter." To build a frequency divider to slow down the pulse rate of the UP board oscillator will require many flip-flop stages. Refer to Figure 5-16.

The equation for the pulse rate is:

$$Qn \text{ rate} = \frac{25,175,000}{2^{n+1}}$$

where n = output number

Connecting an LED to Q17 or Q18 would result in the LED appearing to be always on. Blinking is not visible if the PPS rate is above 30. Connecting an LED to Q19 results in a visible blinking LED.

Grouping any four adjacent outputs creates a MOD 16 counter. Each group counts at a slower rate:

Q0, Q1, Q2, Q3 is a MOD 16 counter with a clock rate of 25,175,000 PPS from the oscillator.

Q1, Q2, Q3, Q4 is a MOD 16 counter with a clock rate of 12,587,000 PPS from output Q0.

Q2, Q3, Q4, Q5 is a MOD 16 counter with a clock rate of 6,293,750 PPS from output Q1.

Q3, Q4, Q5, Q6 is a MOD 16 counter with a clock rate of 3,146,875 PPS from output Q2.

...

Q21, Q22, Q23, Q24 is a MOD 16 counter with a clock rate of 12 PPS from output Q20.

To have a visible count sequence from 0 to 15, on a set of four LEDs, the clock rate must be less than 30 PPS. Connecting LEDs to Q21, Q22, Q23, and Q24 would produce a visible count sequence.

POWERPOINT PRESENTATION

Use PowerPoint to view the Lab 05 slide presentation.

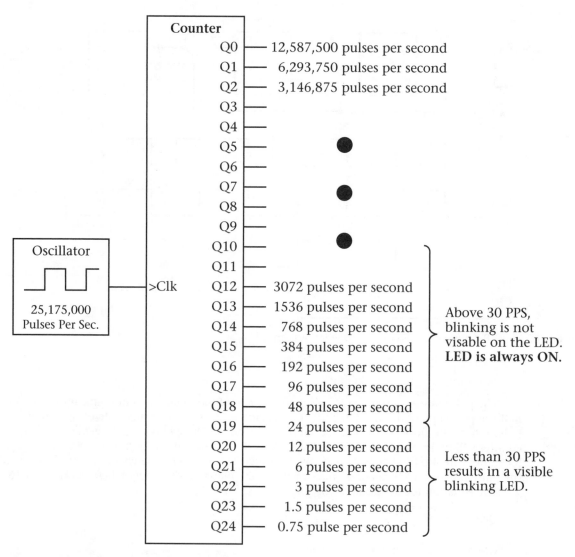

Figure 5-16 UP Oscillator Frequency Divider

LAB WORK PROCEDURE

Design a Random Number Generator System.

To demonstrate the operation of various types of counter systems, a random number generator system will be constructed. A random number generator system is made up of a BCD counter, a 4-bit latch, and a 32-bit frequency divider. The system will generate a random number between 0 and 9 when a pushbutton is pressed. The number will appear on a set of four LEDs.

Add a VHDL BCD Counter and Latch to the System Diagram

You will draw a diagram with a BCD counter and a 4-bit latch that will be part of the random number generator system.

1. A VHDL BCD counter file on the CD-ROM needs to be placed onto a blank disk. Begin by creating a folder (directory) on your blank disk named **Labs**. Copy the file **countbcd.vhd** from the CD-ROM to folder **Labs** on your disk.
   As you have seen in Lab 2, to create a VHDL design you must create a text file to enter the VHDL code. The text file is converted to a symbol that can be displayed in the Graphic Editor. The Graphic Editor is used to connect input and output symbols.

Effectively, your VHDL code creates a new symbol in the Altera symbol library. The VHDL text file representing the counter has been copied from the CD-ROM. You need only create the symbol for the VHDL counter.

2. Start the MAX+PLUS II program. Open the file **countbcd.vhd**. Use the section "Lab Work Procedure: VHDL Vending Machine" in Lab 2 as a guide. Review steps 1 through 5. These steps are not required because the VHDL file was copied from the CD-ROM. You need to open the VHDL file and complete step 6. Step 6 creates the symbol. The text file that represents the "VHDL BCD counter" has been converted into a symbol that can be displayed in the Graphic Editor. The Graphic Editor will be used to connect the latch and input/output symbols.

3. You will draw a diagram for the "BCD counter" and the "latch." Use the Lab Work Procedure in Lab 1 as a guide, and complete steps 1 through 10. Use the project name **Lab5**. The 4-bit latch symbol in the "mega_lpm" library is called **lpm_latch**. The Graphic Editor file for the BCD counter should resemble Figure 5-17.

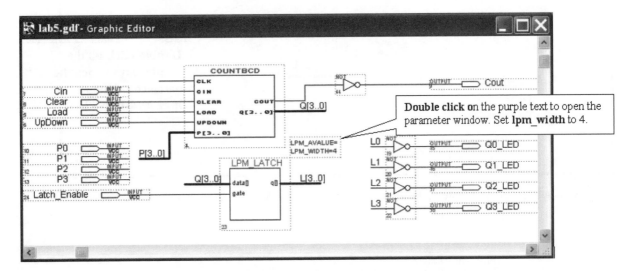

Figure 5-17 Altera "Graphic Editor": Random Number Generator System: BCD Counter and Latch Stages

Add a 32-Bit Frequency Divider to the System Diagram

You will add the 32-bit frequency divider to the diagram for the random number generator system.

1. Use the Lab Work Procedure in Lab 1 as a guide, and complete steps 1 through 12. The 32-bit counter symbol in the "mega_lpm" library is called **lpm_counter**. Refer to Lab 3 section Lab Work Procedure: Build and Test a Switch Debounce System, steps 1 through 5 for assistance regarding connecting the "lpm_counter." The Graphic Editor file should resemble Figure 5-18.

The diagram for the random number generator system is complete. The UP board can now be programmed.

1. Lab 1 describes all the steps that are required to program a UP board with the "random number generator system." Use the Lab Work Procedure in Lab 1 as a guide, and complete steps 13 through 17. Ignore any compiler warning and information messages.

MAX and FLEX IC Considerations

MAX IC Design:
- Connect outputs **Q_LED** and **Cout** to a UP board **MAX LEDs**.
- Connect inputs **Cin**, **Clear**, **Load**, **UpDown**, **P0**, **P1**, **P2**, and **P3** to the UP board **MAX switches**.
- Connect input **Latch_Enable** to the UP board **MAX pushbutton P9** or **P10**.
- Assign input **Oscillator** to pin number **83**. A PCB trace connects the 25,175,000 PPS oscillator to pin 83 of the MAX IC. Refer to Lab 1, step 13_{FLEX} for assistance in assigning an input symbol to a pin number.

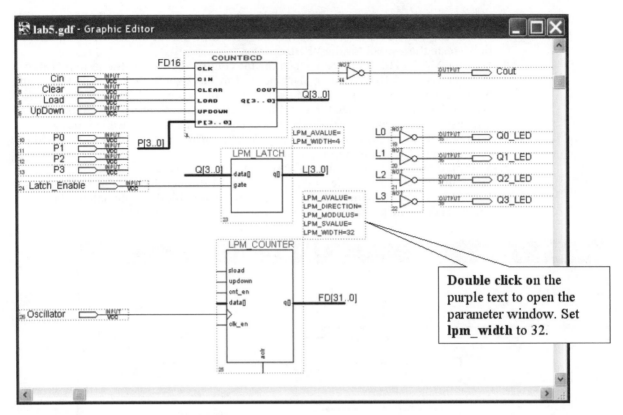

Figure 5-18 Altera "Graphic Editor": Random Number Generator System: Complete System Diagram

FLEX IC Design:
- Assign outputs **Q_LED** and **Cout** to a UP board **FLEX LEDs**. Remember to turn off unused LEDs by connecting them to Vcc.
- Assign inputs **Cin, Clear, Load, UpDown, P0, P1, P2,** and **P3** to the UP board **FLEX switches**.
- Assign input **Latch_Enable** to **pin 28** or **pin 29** to connect a FLEX pushbutton.
- Assign input **Oscillator** to pin number **91**. A PCB trace connects the 25,175,000 PPS oscillator to pin 91 of the FLEX IC.

2. At this point you have drawn the diagram, programmed the IC, and connected all switches and LEDs. You are ready to test the operation of the random number generator system.

Lab Exercises and Questions

This section contains lab exercises that can be performed on the UP board and questions that can be answered at home. Ask your instructor which exercises to perform and which questions to answer.

Lab Exercises

Exercise 1: Isolate and test the BCD counter section of the random number generator system.

You will slow down the pulse rate of the BCD counter and test the effects of various control inputs.

a. Change the pulse rate of the BCD counter clock to **3 PPS**. Currently, **FD16** is used and it is too fast. You will need to use the introductory section of the lab to figure out this change. Recompile and reprogram the Altera IC. Set switches **Cin** and **Clear** to 1. Set switches **Load, UpDown, P3, P2, P1,** and **P0** to 0. Do not press the **Latch_Enable**

pushbutton. Explain the LED response at **Q3 ... Q0** and **Cout**.
b. Set **Cin** to **0** and then back to **1**. Explain the **LED** response when "Cin" = 0.
c. Set **Clear** to **0** and then back to **1**. Explain the **LED** response when "Clear" = 0.
d. Set **Load** to **1** and **P3, P2, P1**, and **P0** to **0110**, respectively. Set **Load** to **0**. Explain the **LED** response.
e. Set **UpDown** to **1**. Explain the **LED** response.
Change the BCD counter clock name back to **FD16**. Recompile and reprogram the Altera IC.

Exercise 2: Test the complete random number generator system.

a. Set switches **Cin** and **Clear** to **1**. Set switches **Load, UpDown, P3, P2, P1**, and **P0** to **0**. Do not press the **Latch_Enable** pushbutton. Calculate the **pulse rate** of the **BCD counter** clock connected to "FD16." Calculate the pulse rate at output **Q3_LED**. Use these calculations to explain why "Q3_LED" is blinking.
b. Press the **Latch_Enable** pushbutton and hold it down (logic 0) for 1 or 2 seconds. Release the **Latch_Enable** pushbutton (logic 1) for 1 or 2 seconds. Repeat the press, hold, and release sequence four or five times. Explain the **LED** response at output **Q3, Q2, Q1**, and **Q0**.
c. Why is the "Cout" LED flickering when the "Latch_Enable" button is being held down?
d. Explain how the system works as a random number generator system.

Exercise 3: Use the Altera simulator to test the operation of the random number generator system.

(This exercise can be done at home.)

Use the Max+Plus II Graphic Editor to make changes to the **lab5.gdf** file for the random number generator system. The changes will simplify the system and allow you to use the simulator to test its operation.

1. Delete the input symbols **Cin, Clear, Load, UpDown, P3, P2, P1**, and **P0**.
2. Connect a **Vcc** symbol to **Cin** and **Clear**.
3. Connect a **Gnd** symbol to **Load, UpDown, P3, P2, P1**, and **P0**.
4. Change the name of the **COUNTBCD Clk** input from **FD16** to **FD0**.
5. Add a second set of four **NOT** gates to **Q0_LED, Q1_LED, Q2_LED**, and **Q3_LED**. The changes are shown in Figure 5-19.

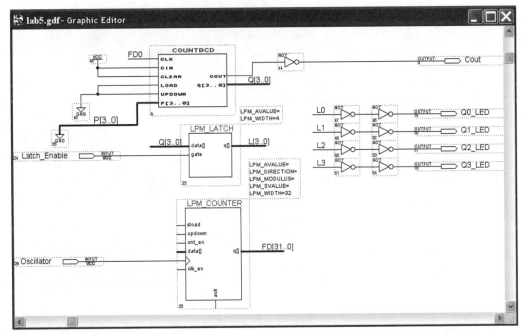

Figure 5-19 Altera "Graphic Editor": Random Number Generator System: Diagram Used for Simulation

Use Appendix F as a guide to run a simulation of the random number generator system you have just created. The changes will give warnings when you use "Save, Compile & Simulate." These warnings can be ignored. The "Waveform Editor" window for the simulation should resemble Figure 5-20.

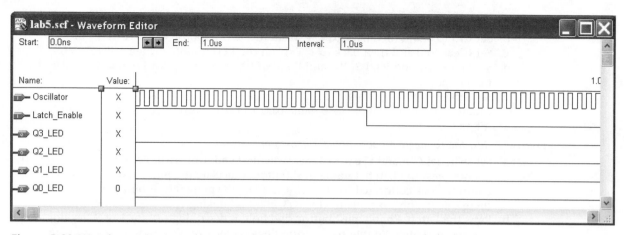

Figure 5-20 Waveform Editor Window for Random Number Generator System

Set the **End time** to 1.0us. The "Oscillator" uses an "Overwrite clock period" = 20.0ns, and the "Latch_Enable" uses an "Overwrite clock period" = 1.0us. If the "Clock Period" entry box is grayed out (not selectable), then you should exit the "Overwrite Clock" window, select the **Options** menu, and remove the check mark beside the menu item "Snap to Grid." Return to the "Overwrite Clock" window and you will be able to set the clock period. If you notice that the "Latch_Enable" waveform starts at 0 then changes to 1, then you will have to invert the waveform. To invert the "Latch_Enable" waveform, right click on it and then select **Overwrite** then **Invert**.

   a. Run the simulation and explain the output response you observe at the "Q" outputs.
   b. Group the **Q** outputs into a single node and run the simulation again. You may need to increase the scale (use the magnify glass with a + inside) to view the response of the simulation. Explain the output response you observe at the grouped node for output "Q."
   c. Why must you change the "COUNTBCD" "Clk" input from "FD16" to "FD0"?
   d. Why invert the "Latch_Enable" input? Ungroup the "Q" output node and run the simulation with the "Latch_Enable" input *not* inverted.

Exercise 4: Design a random number generator with a range of 0 to 99.

Cascade a second BCD counter and expand the latch to create a "0" to "99" random number generator. Display the BCD numbers on eight UP board LEDs.

Exercise 5: Design a VHDL stopwatch.

(This exercise can be assigned as a project.)

The stopwatch will have a start button, a lap button, a stop button, and an LED binary display. Refer to Figure 5-21.

   • *Start Button*: Press and hold down (logic 0) resets the stopwatch to zero and blocks counting. Release (logic 1) makes the watch display the time advancing from 0 to 59 seconds.
   • *Lap Button*: Press and hold down (logic 0) freezes the current time on the display while the watch continues to track the advancing time in the background. Release (logic 1) makes the watch revert back to displaying the advancing time.
   • *Stop Button:* Press and hold (logic 0) stops the watch. Release (logic 1) keeps the watch stopped. The display shows the stop time. The start button must be used to restart the stopwatch.

Exercise 6: Design a parking garage system.

(This exercise can be assigned as a project.)

Design a parking garage system that will hold a maximum of nine cars. It uses two sensors. One sensor detects cars coming into the parking garage and the other detects the cars leaving the garage. A gate is lowered to block cars from entering the parking garage when the parking lot is full. Refer to Figure 5-22.

Exercise 7: Design a elevator control system.

(This exercise can be assigned as a project.)

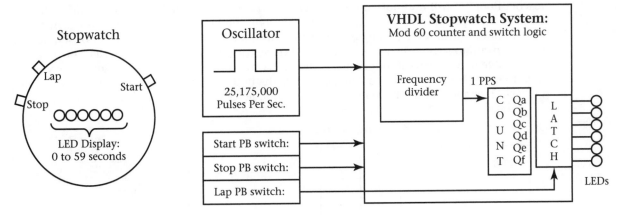

Figure 5-21 Stopwatch System

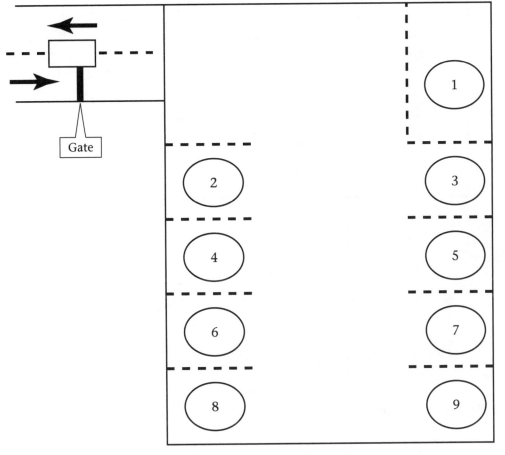

Floor Plan

Figure 5-22 Parking Garage System

Inside the elevator there will be a control panel made up of four buttons and four LEDs. They will be labeled floor 1, floor 2, floor 3, and floor 4. Press a button and the elevator will move to the floor and stop. Let's assume the elevator is on floor 1 and no one is inside the elevator pressing buttons. Thus, all switches are 0 and the LED for floor 1 is lit. A person enters the elevator and presses button #3 (switch = 1). The lit LED will move from floor 1 to floor 2 then to floor 3 simulating the motion of the elevator. There will only be one lit LED. When the elevator reaches floor 3 it will not move until another button is pressed (switch #3 = 0 and a different switch = 1). A person now enters the elevator on floor number 3 and presses button #2 (switch = 1). The elevator will reverse its motion and move to floor 2 immediately. The elevator should not move forward to floor 4 then reverse itself and come back down to floor 2. The elevator should also be able to reverse itself immediately when it has to move up to a new floor after it has been used to move down. Your design will *not* include the buttons that are located outside the elevator on each floor and will not include the elevator door. Refer to Figure 5-23.

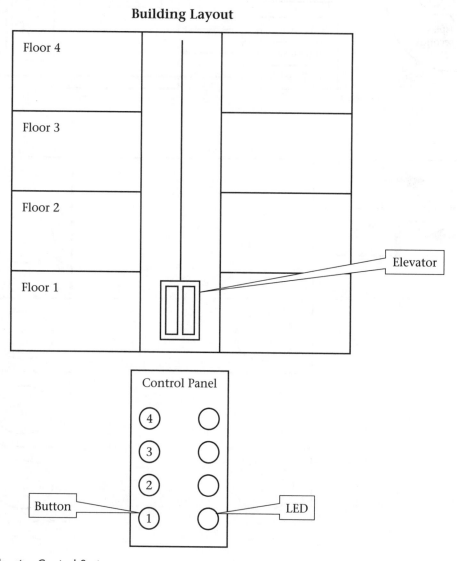

Figure 5-23 Elevator Control System

Lab Questions

1. Draw the diagram for a Mod 16 ripple up counter. Draw the waveform diagram and fill in a count table.
2. Draw the diagram for a Mod 8 ripple down counter. Draw the waveform diagram and fill in a count table.
3. Draw the diagram for a Mod 4 combination ripple up/down counter.
4. Draw the diagram for a Mod 8 parallel UP counter.
5. Design a VHDL switch debounce system. There are three main sections to the debounce system. Starting at the debounce output, they are:
   a. A four-input AND gate
   b. A four-bit shift register
   c. A 17-stage counter (Q0 to Q16) clocked at 25,175,000 PPS

To design a VHDL debounce system, you must create a single VHDL file that integrates the three main sections. The VHDL code for the 4-input AND gate is:

IF SHIFT4(3 DOWNTO 0)="1111" THEN

    DEBOUNCE <= '1';

ELSE

    DEBOUNCE <= '0';

END IF;

CONCLUSIONS

- A JK flip-flop is an improved SC flip-flop. The ambiguous mode has been replaced with the toggle mode.
- JK flip-flops connected to each other operating in the toggle mode create a binary counter system.
- Connect Q to the clock of the next flip-flop creates an up counter. Connect $\overline{Q}$ to the clock of the next flip-flop creates a down counter.
- Cin and Cout are used to cascade counter symbols.
- BCD is short for binary coded decimal. It is a number system that is used to represent decimal numbers in binary.
- Counters can be used to divide the frequency of a high-speed clock.
- The pulse rate must be less than 30 PPS to be visible on an LED.
- A random number generator can be created with a high-speed counter and a latch.

NOTE: *Lab 6 will use the random number generator. Do not delete any Lab 5 files in folder "Labs." Store this disk safely until it's needed.*

Lab 6

Digital Display Decoder System

Equipment   Altera UP board
Book CD-ROM
Spool of 24 AWG wire
Wire cutters
Floppy Disk with Lab 5 files

Objectives   Upon completion of this lab, you should be able to:
- Operate a digital display
- Operate digital display decoders.
- Convert decimal numbers to hexadecimal numbers.
- Convert hexadecimal numbers to decimal numbers.

INTRODUCTORY INFORMATION

Up until now, the systems you have designed utilized individual LEDs to display binary output data. An understanding of the binary number system was required in order to interpret the output response. In this lab you will use digital displays similar to the ones you see on digital clock radios and other consumer electronic products. The system response will be much easier to interpret.

Digital Displays

Digital displays are called seven-segment displays. The seven-segments are rectangular lenses that house seven LEDs. The lenses are arranged to make up the pattern of the number 8. Turning on or off combinations of LEDs will allow you to display the numbers 0 through 9. Refer to Figure 6-1.

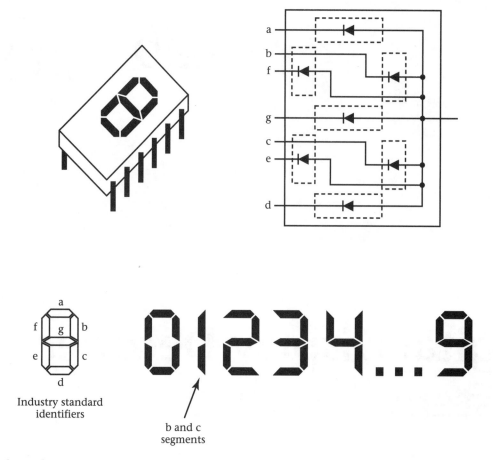

Figure 6-1 Seven-Segment LED Display

There are two types of seven-segment LED digital display packages. They are called the **common anode** display and the **common cathode** display. The common anode display connects all the anodes to a common connection line on the display package. It is an active low display package. Active low refers to the fact that a logic 0 at the segment input lights up the LED. The common cathode display package, on the other hand, has all the diodes reversed. The common cathode line is connected to ground (logic 0) and a segment lights up when it is connected to 5 volts (logic 1). It is an active high display package. Refer to Figure 6-2.

The Altera UP board uses active low, common anode displays.

Basic Operation of a Seven-Segment Decoder

A seven-segment decoder is a digital device that is used to control a seven-segment display. Refer to Figure 6-3. The decoder has seven outputs connected to segments a, b, … g. The bubble at the output signifies that they are active low. An active low output is asserted with a "0." A "0" is required to light a segment on a common anode digital display. The decoder has three control inputs and one control output. Their function is described later. The basic operation of a seven-sgement decoder is simple. A 4-bit number is applied to inputs **DCBA**, the device decodes the number and sends out the appropriate combinations of 1s and 0s to activate the LEDs and display the number. Input D is the most significant bit and A is the least significant bit. Refer to Figure 6-4.

Seven-Segment Decoder: Control Inputs LT and BI

LT is the Lamp Test input. It is active low. When asserted it overrides DCBA and sends signals to the digital display that will turn *on* all the segments. "LT" is used to check if the display has any burned-out LEDs. If a digital system consists of multiple digital displays, this control feature can

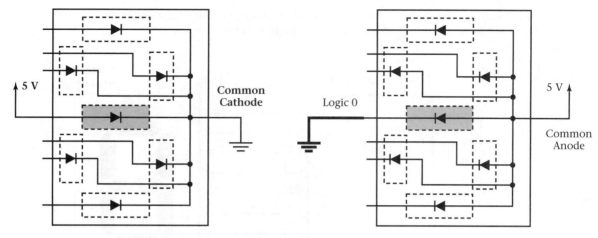

Figure 6-2 Common Cathode and Common Anode Displays

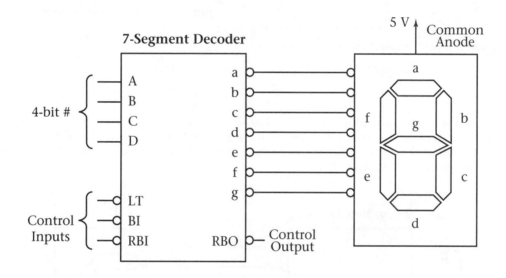

Note: The diode resistors have been omitted to simplify the diagram.

Figure 6-3 Seven-Segment Decoder

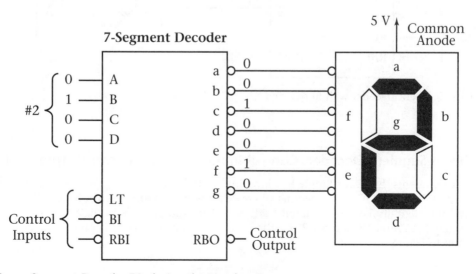

Figure 6-4 Seven-Segment Decoder Displaying the Number 2

be used by a maintenance technician to easily inspect the state of the entire display. If this control feature is not desired, connect LT to logic 1. Refer to Figure 6-5.

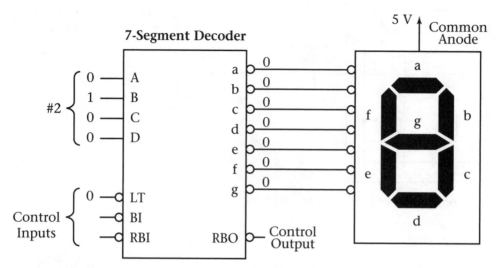

Figure 6-5 Seven-Segment Decoder with LT Asserted

BI is the Blanking Input. It is active low. When asserted it overrides DCBA and sends signals to the digital display that will turn *off* all the segments. BI can be used like a safety valve to quickly shut down the display whenever necessary. If this control feature is not desired, connect BI to logic 1. Refer to Figure 6-6.

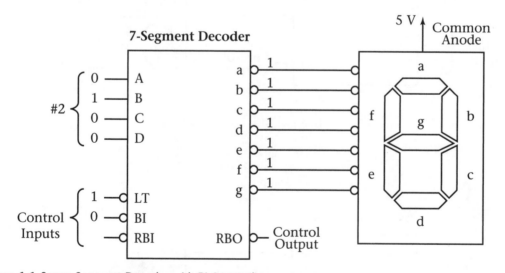

Figure 6-6 Seven-Segment Decoder with BI Asserted

Seven-Segment Decoder: Control Input RBI And Control Output RBO

RBI represents the words Ripple Blanking Input. **RBO** represents the words Ripple Blanking Output. RBI and RBO work together to create a ripple blanking control feature that allows multiple digit displays to blank out leading 0s. For example, if a six-digit display is used to display the four-digit number 2047, the visual appearance of 2047 is preferable over 002047. Refer to Figure 6-7.

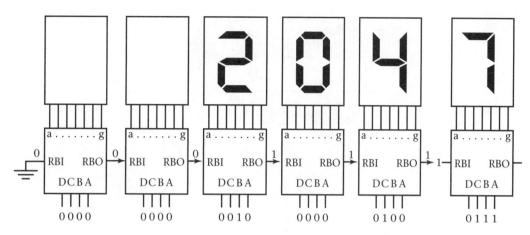

Figure 6-7 Six-Digit Display Displaying a Four-Digit Number with Leading 0s Blanked

On any display stage, if "RBI" and "DCBA" inputs are all at logic "0" the display will blank and the "RBO" will be at logic "0." Chaining "RBO" to "RBI" will signal the next least significant stage that the leading zero has been blanked. This stage can also blank the next leading zero if necessary. If any input "RBI," "D," "C," "B," or "A" are at logic "1," the display will not blank the number and "RBO" will be set to logic "1." The decoder at the far left (MSD) has its "RBI" connected to "0" and will never be able to display the number "0." The decoder at the far right (LSD) has its "RBI" connected to "1" and will always display the number "0." Connecting the LSD stage's "RBI" to the "RBO" of the adjacent stage would cause the entire display to be off when the six-digit number was "000000." Imagine how frustrating it would be if the display system was used in a calculator and the "Power On" button was pressed.

Hexadecimal Numbers (HEX Numbers)

As you have seen, the seven-segment decoder receives a 4-bit number at inputs "DCBA;" it decodes the number and displays it on the digital display. The question is: What is displayed when the number at DCBA is larger than 9? The 4-bit binary numbers larger than 9 are: "1010," "1011," "1100," "1101," "1110," and "1111." They represent the decimal numbers "10," "11," "12," "13," "14," and "15." How does a two-digit number get displayed on a single-digit display? Some seven-segment decoders are called BCD to seven-segment decoders. This implies that the device can only be used with the numbers 0 through 9 at inputs "DCBA." Other seven-segment decoders are called hexadecimal to seven-segment decoders. These decoders can display all 4-bit number combinations at inputs DCBA, including the numbers larger than 9. To understand how they work you must know how the **hexadecimal number system** works.

The HEX number system is a base 16 number system. It has 16 numerals. They are 0, 1, 2, 3, 4, 5, 6, 7, 8, 9, A, B, C, D, E, and F. HEX uses the 10 numerals of the decimal number system and the first 6 letters of the alphabet. Letters A, B, C, D, E, and F are used as numbers because the shape of the six symbols and the numerical order of the symbols (examples: A is smaller than B; F is larger than C; etc.) are already known.

Counting in HEX can be explained by looking at a car with a three-digit HEX odometer. It is made up of three plastic discs. Each disc is labeled with all the numerals of the number system. The discs spin as the car is traveling forward. The driver of the car views the odometer by looking at the front of the discs through a rectangular opening. Refer to Figure 6-8.

The HEX number system is identical to the decimal number system from 0 to 9. At 9 the HEX number system increments forward using the HEX numerals A though F. At 19 the HEX number system increments forward by using HEX numerals 1A through 1F. These transitions from 9 to A, or 19 to 1A, or 29 to 2A make this number system unique and a bit confusing.

The main benefit of the HEX number system is to represent binary numbers in a compressed format. Digital systems like calculators, CD players, or hard disk drives are designed to process binary data. The binary data can be 8, 12, 16, 32, or 64 bits in length. To design, maintain, or repair these systems requires diagnostic equipment or diagnostic software. Diagnostic equipment allows you to view the binary data that is processed by the system. Working with 16-, 32- or 64-bit data codes in binary is very awkward. A large binary number looks like a sea of 1s and 0s. The

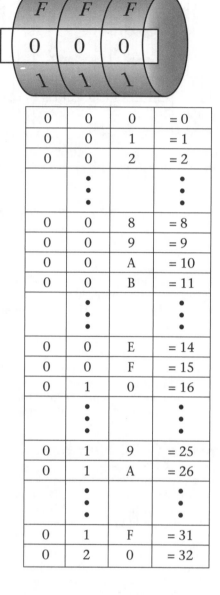

0	0	0	= 0
0	0	1	= 1
0	0	2	= 2
	•		•
	•		•
	•		•
0	0	8	= 8
0	0	9	= 9
0	0	A	= 10
0	0	B	= 11
	•		•
	•		•
	•		•
0	0	E	= 14
0	0	F	= 15
0	1	0	= 16
	•		•
	•		•
	•		•
0	1	9	= 25
0	1	A	= 26
	•		•
	•		•
	•		•
0	1	F	= 31
0	2	0	= 32

Figure 6-8 Hexadecimal Odometer

same data displayed in HEX is compact and easier for you to work with. For example, here is a 32-bit number displayed in binary: 10111100010101001011110001010100; here is the same number in HEX: BC54BC54. If you had to memorize one of these numbers, which one would you prefer?

A 4-bit binary PWC shows how a single HEX digit compresses a 4-bit binary number. Refer to Figure 6-9. To convert a binary number to HEX, all you need to do is start at the least significant bit, group the bits into blocks of four, and convert each block of 4 bits to HEX.

EXAMPLE: Convert the following binary number to HEX: 1110111100.

Group bits into blocks of four:
11   1011   1100
Convert each block to HEX.
11 = 0011 = 3   1011 = B (11 decimal)   1100 = C (12 decimal)
Final answer: 1110111100 binary = 3BC HEX

To convert a HEX number to binary, all you need to do is convert each HEX numeral to a 4-bit binary number.

EXAMPLE: Convert the following HEX number to binary: 5AE.

Bin. PWC 8	4	2	1	Decimal	HEX
0	0	0	0	0	0
0	0	0	1	1	1
0	0	1	0	2	2
0	0	1	1	3	3
0	1	0	0	4	4
0	1	0	1	5	5
0	1	1	0	6	6
0	1	1	1	7	7
1	0	0	0	8	8
1	0	0	1	9	9
1	0	1	0	10	A
1	0	1	1	11	B
1	1	0	0	12	C
1	1	0	1	13	D
1	1	1	0	14	E
1	1	1	1	15	F

Figure 6-9 4-Bit Binary PWC with Decimal and HEX Equivalents

5 = 0101   A = 1010   E = 1110
Final answer: 5AE = 0101 1010 1110 = 10110101110 binary

VHDL Seven-Segment Decoder

The VHDL code for seven-segment decoder, with active low outputs, is shown in Figure 6-10. The architecture uses a vector statement **SIGNAL input: STD_LOGIC_VECTOR (3 DOWNTO 0);** to define a storage array. The storage array is called **input**, and **(3 DOWNTO 0)** implies it stores 4 bits of data. The statements **input (0) <= Ain;**, **input (1) <= Bin;**, **input (2) <= Cin;**, and **input (3) <= Din;** assign the input variables to the storage array. These four inputs represent the binary number that is to be displayed.

The architecture uses a vector statement **SIGNAL output: STD_LOGIC_VECTOR (6 DOWNTO 0);** to define a storage array. The storage array is called **output**, and **(6 DOWNTO 0)** implies it stores 7 bits of data. The statements **a <= output(6);**, **b <= output(5);**, **c <= output(4);**, **d <= output(3);**, **e <= output(2);**, **f <= output(1);**, and **g <= output(0);** assign the output array to the output variables *a, b, c, d, e, f,* and *g*. These seven outputs represent the display segments that are used to display the number.

The code for the Ripple Blanking feature is included in the statements:
```
IF (input = "0000" and nRBI ='0') THEN  — Blank leading 0
        output   <=      "1111111";
        nRBO   <=      '0';
ELSIF (input = "0000" and nRBI = '1') THEN – display 0
        output   <=      "0000001";
        nRBO   <=      '1';
```
If the number at Ain, Bin, Cin, and Din is "0000" and RBI is "0," then "1111111" is assigned to the output segments (a, b, ... , g). This will turn off the active low LEDs of the digital display. However, if the RBI is "1," then "0000001" is assigned to the output all segments (a, b, ... , g). This will display the number 0 on the digital display.

The **Case** statement is a **conditional statement** that is used as an alternative to using the "IF" statement. The expression **input** (representing **DCBA**) at the top of the "Case" statement is evaluated and compared with the expressions following each **WHEN**. The assign **output**

```
LIBRARY ieee;
USE ieee.std_logic_1164.ALL;
ENTITY seven_seg IS
        PORT(
                Ain, Bin, Cin, Din, nLT, nBI,  nRBI   : IN   STD_LOGIC;
                a, b, c, d, e, f, g, nRBO             : OUT  STD_LOGIC);
END seven_seg;

ARCHITECTURE decoder OF seven_seg IS
        SIGNAL input: STD_LOGIC_VECTOR (3 DOWNTO 0);
        SIGNAL output: STD_LOGIC_VECTOR (6 DOWNTO 0);
BEGIN
-- Concurrent signal assignment
input (0) <= Ain;
input (1) <= Bin;
input (2) <= Cin;
input (3) <= Din;

display:
PROCESS (input, nRBI, nBI, nLT)
BEGIN
        --  If Statement
                IF (input = "0000" and nRBI ='0') THEN   -- Blank leading 0
                        output  <=       "1111111";
                        nRBO    <=       '0';
                ELSIF (input = "0000" and nRBI = '1') THEN -- display 0
                        output  <=       "0000001";
                        nRBO    <=       '1';
                ELSE
                        CASE input IS
                                WHEN "0001"   =>      output <=    "1001111"; --display 1
                                WHEN "0010"   =>      output <=    "0010010"; --display 2
                                WHEN "0011"   =>      output <=    "0000110"; --display 3
                                WHEN "0100"   =>      output <=    "1001100"; --display 4
                                WHEN "0101"   =>      output <=    "0100100"; --display 5
                                WHEN "0110"   =>      output <=    "0100000"; --display 6
                                WHEN "0111"   =>      output <=    "0001111"; --display 7
                                WHEN "1000"   =>      output <=    "0000000"; --display 8
                                WHEN "1001"   =>      output <=    "0000100"; --display 9
                                WHEN "1010"   =>      output <=    "0001000"; --display A
                                WHEN "1011"   =>      output <=    "1100000"; --display B
                                WHEN "1100"   =>      output <=    "0110001"; --display C
                                WHEN "1101"   =>      output <=    "1000010"; --display D
                                WHEN "1110"   =>      output <=    "0110000"; --display E
                                WHEN "1111"   =>      output <=    "0111000"; --display F
                                WHEN others   =>      output <=    "1111111"; --blank

                        END CASE;
                        nRBO    <=       '1';
                END IF;
                IF (nBI='0') THEN       -- override DCBA and blank the display
                        output  <=       "1111111";
                        END IF;
                IF (nLT='0') THEN       -- override DCBA and lamp test.
                        output  <=       "0000000";
                        END IF;

        a    <=      output(6);
        b    <=      output(5);
        c    <=      output(4);
        d    <=      output(3);
        e    <=      output(2);
        f    <=      output(1);
        g    <=      output(0);
END PROCESS display;
END decoder;
```

input:

(3)	(2)	(1)	(0)
D	C	B	A

output:

(6)	(5)	(4)	(3)	(2)	(1)	(0)
a	b	c	d	e	f	g

Figure 6-10 VHDL code for a seven-segment decoder.

statement (segment **LEDs**) within the matching **WHEN** branch is executed and then control jumps to the statement following **END CASE**. The case statement assigns a 7-bit number to the output segments based on an input condition at Ain, Bin, Cin, and Din. For example, the statement **WHEN "0001" => output <= "1001111"; —display 1** implies that **1001111** is assigned to the output segments when the input (Ain, Bin, Cin, and Din) is **0001**. The 7-bit number **1001111** will light segments b and c to display the number 1 when number 1 is at the input. This is the correct response. The remaining case statements provide 7-bit numbers that are used to display the remaining digits. It is worth noting the difference between displaying the number "6" and the hexadecimal number "B." The number "6" lights the "a" segment, whereas the hexadecimal number "B" does not. This displays a lowercase "B" because a capital "B" would be confused with the number "8."

The code for the **Blanking** and **Lamp Test** features are included in the statements:

```
IF (nBI='0') THEN        — override DCBA and blank the display
        output  <=      "1111111";
        END IF;
IF (nLT='0') THEN        — override DCBA and lamp test.
        output  <=      "0000000";
        END IF;
```

Assigning "1111111" to the output segments turns off all LEDs, and assigning "0000000" to the output segments turns on all LEDs. Both are the correct response.

POWERPOINT PRESENTATION

Use PowerPoint to view the Lab 06 slide presentation.

LAB WORK PROCEDURE

Add a Digital Display to the Random Number Generator System of Lab 5

In Lab 5 you designed a random number generator system. The system displayed a random number between 0 and 9 when a pushbutton was pressed. The number was displayed on a set of four LEDs. You will add a digital display to the system.

VHDL Seven-Segment Display Decoder:

In this step, you will add a **VHDL display decoder symbol** to the **Altera symbol library**.

1. A VHDL display decoder file on the CD-ROM needs to be placed onto the disk that stores the Lab 5 files. Copy the file **seven_seg.vhd** from the CD-ROM to folder **Labs** on your disk.

   As you have seen in Lab 2, to create a VHDL design you must create a text file to enter the VHDL code. The text file is converted to a symbol that can be displayed in the Graphic Editor. The Graphic Editor is used to connect input and output symbols. Effectively, your VHDL code creates a new symbol in the Altera symbol library. The VHDL text file representing the display decoder has been copied from the CD-ROM. You need only create a symbol for the VHDL display decoder symbol.

2. Start the **MAX+PLUS II** program. Open the file **seven_seg.vhd**. Use the section "Lab Work Procedure: VHDL Vending Machine" in Lab 2 as a guide. Review steps 1 through 5. These steps are not needed because the VHDL file was copied from the CD-ROM. You only need to open the VHDL file and complete step 6. Step 6 creates the symbol. Ignore any warning message.

Add the Display Decoder to the Random Number Generator System

The text file that represents your VHDL display decoder has been converted into a symbol that can be displayed in the Graphic Editor. You will open the file for the random number generator system from Lab 5 and add the VHDL display decoder symbol.

1. Open the file **lab5.gdf**. Use the Lab Work Procedure in Lab 1 as a guide, and complete steps 1 through 12. You will need to make several changes:

   a. Delete **Output** symbols and **NOT** gates for **Q0, Q1, Q2, Q3,** and **Cout**.

b. Delete the **Input** symbols **P0**, **P1**, **P2**, and **P3** and connect the input bus to ground. The ground symbol in the "prim" library is called **gnd**.
c. Delete the **Input** symbols **Load** and **UpDown** and connect the inputs to ground.
d. Delete the **Input** symbols **Cin** and **Clear** and connect the inputs to **Logic 1** or **Vcc**. The **Vcc** symbol in the "prim" library is called **vcc**.
e. Add the VHDL display decoder symbol.

The Graphic Editor file for the system should resemble Figure 6-11.

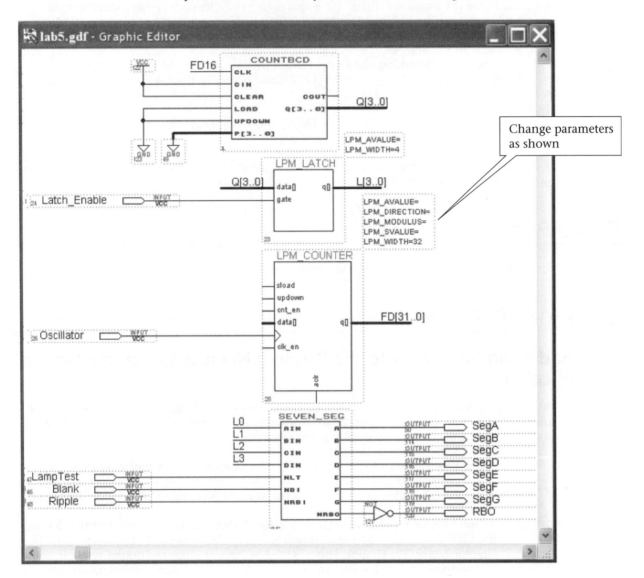

Figure 6-11 Altera "Graphic Editor": Random Number Generator System with a Digital Display

2. The UP board can now be programmed. Lab 1 describes all the steps that are required to program a UP board with the random number generator system. Use the Lab Work Procedure in Lab 1 as a guide, and complete steps 13 through 17. Ignore any compiler warning and information messages.

MAX and FLEX IC Considerations

MAX IC Design:
- Connect **LampTest**, **Blank**, **Ripple** to UP board **MAX** switches.
- Connect **Latch_Enable** to UP board **MAX** pushbutton **P9** or **P10**.

- Assign input **Oscillator** to pin number **83**. A PCB trace connects the 25,175,000 PPS oscillator to pin 83 of the MAX IC. Refer to Lab 1, step 13$_{FLEX}$ for assistance in assigning an input symbol to a pin number.
- Assign pin numbers to output symbols **SegA**, **SegB**, … **SegG** to connect the MAX digital display to the display decoder. The MAX IC has a pair of digital displays. PCB traces connect the active low LED displays to the IC. The pin numbers for each digital display are shown in Figure 6-12. Refer to Lab 1, step 14$_{FLEX}$ for assistance.
- Assign output symbol **RBO** to the decimal point LED.
- You need to turn off all of the other unused LEDs in order to block the MAX+Plus II software from randomly routing logic signals through these pins and turning them on. Having unused LEDs inadvertently turn on is a visual distraction. Active low LEDs can be turned off with a connection to logic 1. Refer to Lab 1, step 14$_{FLEX}$ for assistance.
- You are ready to test the operation of the random number generator system.

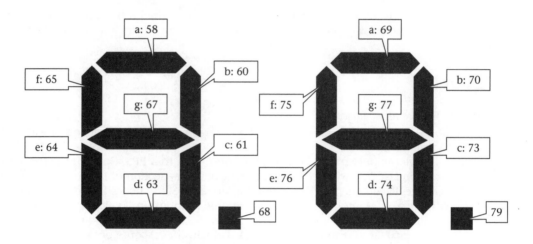

MAX Digital Display

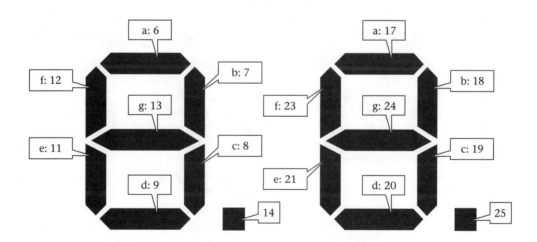

FLEX Digital Display

Figure 6-12 Pin Numbers for MAX and FLEX Digital Displays

FLEX IC Design:
- Assign **LampTest**, **Blank**, **Ripple** to connect to UP board **FLEX** switches.
- Assign **Latch_Enable** to **pin 28** or **pin 29** to connect a FLEX pushbutton.

- Assign input **Oscillator** to pin number **91**. A PCB trace connects the 25,175,000 PPS oscillator to pin 91 of the FLEX IC.
- Assign pin numbers to output symbols **SegA, SegB, … SegG** to connect the FLEX digital display to the display decoder. The FLEX IC has a pair of digital displays. PCB traces connect the active low LED displays to the IC. The pin numbers for each digital display are shown in Figure 6-12. Refer to Lab 1, step 14$_{FLEX}$ for assistance.
- Assign output symbol **RBO** to the decimal point LED.
- Don't forget to turn off all of the other unused LEDs.
- You are ready to test the operation of the random number generator system.

LAB EXERCISES AND QUESTIONS

This section contains lab exercises that can be performed on the UP board and questions that can be answered at home. Ask your instructor which exercises to perform and which questions to answer.

Lab Exercises

Exercise 1: Test the display decoder.

You will slow down the pulse rate of the BCD counter clock and test the effects of various display decoder control inputs.

   a. Change the **pulse rate** of the **BCD counter** clock to **3 PPS**. Currently, FD16 is used and it is too fast. You will need to use the introductory section of Lab 5 to figure out this change. Recompile and reprogram the Altera IC. Set switches **LampTest, Blank**, and **Ripple** to 1. Do not press the **Latch_Enable** pushbutton. Explain the response of the digital display.
   b. Set **LampTest** to 0 and then back to 1. Explain the response of the digital display when **LampTest = 0**.
   c. Set **Blank** to 0 and then back to 1. Explain the response of the digital display when **Blank = 0**.
   d. Set **Ripple** to 0. Watch the digital display cycle through the entire count sequence. Set **Ripple** to 1. Watch the digital display cycle through the entire count sequence. Explain the response of the digital display when **Ripple = 0**. Explain the response of the digital display when **Ripple = 1**. What is different? Explain the response of the **RBO** output LED.

Change the name of the BCD counter clock back to **FD16**. Recompile and reprogram the Altera IC.

Exercise 2: Test a random number generator system.

Set switches **LampTest, Blank**, and **Ripple** to 1. Press the latch-enable pushbutton and hold it down (logic 0) for 1 or 2 seconds. Release the latch-enable pushbutton (logic 1) for 1 or 2 seconds. Repeat the press, hold, and release sequence four or five times. Explain how the system works as a random number generator.

Exercise 3: Add digital displays to a random number generator system with a range of 0 to 99.

Cascade a second BCD counter and expand the latch to create a 0 to 99 random number generator. Display the BCD numbers on two digital displays.

   a. Use the **ripple blanking** feature to blank the leading **0**. Figure out how to connect the **NRBO** outputs and the **NRBI** inputs.
   b. Change the **pulse rate** of the **BCD counter** clock to **3 PPS**. Currently FD16 is used and it is too fast. You will need to use the introductory section of the Lab 5 to figure out this change. Recompile and reprogram the Altera IC. Explain the response of the digital displays.
   c. Disable the ripple blanking feature. Remove the NRBO connections and disable the "NRBI" inputs by directly connecting them to "logic 1." Recompile and reprogram the Altera IC. Explain the response of the digital displays.

Lab Questions

1. Convert each binary number to hexadecimal
   a. 10011011
   b. 11110
   c. 111
   d. 1101101
2. Convert each hexadecimal number to binary
   a. 1C0
   b. 7E
   c. F0D
   d. DAD
3. Each system shown in Figure 6-13 has a connection error. For each system, fill in the table to show what would appear on the digital display.
4. Change the VHDL code shown in Figure 6-10 to create a seven-segment decoder with active high outputs.

Conclusions

- Seven-segment common anode LED displays are active high displays, and common cathode LED displays are active low displays.
- Lamp test is used to check for burned-out LED segments.
- Blanking input is used to turn off the display.
- Ripple blanking input and ripple blanking output are used to turn off leading zeros.
- Hexadecimal number system compresses a 4-bit number to a single digit.
- Hexadecimal uses letters A, B, C, D, E, F as numerals.
- The VHDL display decoder uses a Case statement to assign logic levels to the segment outputs.

(A) The BCD counter is connected backwards to the display decoder. For example, the MSB of the counter is connected to the LSB of the display decoder.

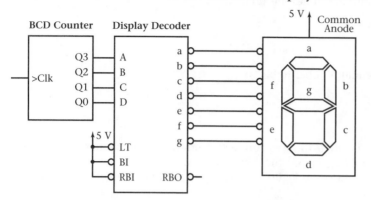

Clk	Q3	Q2	Q1	Q0	D C B A	Display
0	0	0	0	0		
1	0	0	0	1		
2	0	0	1	0		
3	0	0	1	1		
4	0	1	0	0		
5	0	1	0	1		
6	0	1	1	0		
7	0	1	1	1		
8	1	0	0	0		
9	1	0	0	1		

(B) Lamp test is connected to Q3.

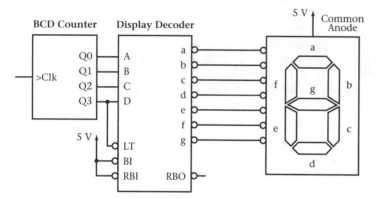

Clk	Q3	Q2	Q1	Q0	D C B A	Display
0	0	0	0	0		
1	0	0	0	1		
2	0	0	1	0		
3	0	0	1	1		
4	0	1	0	0		
5	0	1	0	1		
6	0	1	1	0		
7	0	1	1	1		
8	1	0	0	0		
9	1	0	0	1		

(C) RBI is connected to Q3.

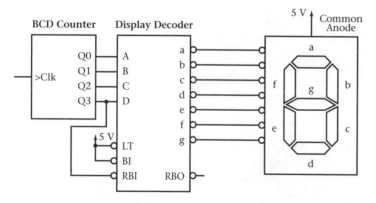

Clk	Q3	Q2	Q1	Q0	D C B A	Display
0	0	0	0	0		
1	0	0	0	1		
2	0	0	1	0		
3	0	0	1	1		
4	0	1	0	0		
5	0	1	0	1		
6	0	1	1	0		
7	0	1	1	1		
8	1	0	0	0		
9	1	0	0	1		

Figure 6-13 Display Decoder Systems for Question 3

Lab 7

"1 of X" Decoder and Encoder Systems

Equipment

Altera UP board
Book CD-ROM
Spool of 24 AWG wire
Wire cutters
Floppy Disk with Lab 3 debounce files

Objectives

Upon completion of this lab, you should be able to:

- Operate a "1 of X" decoder.
- Operate a priority "1 of X" encoder.
- Design and test a keypad encoder system.
- Design and test a two-key storage system.

INTRODUCTORY INFORMATION

As you have seen in Lab 6, a display decoder receives a 4-bit binary number and controls a set of LEDs to display the number. The decoder receives data and generates an output response based on the data received. In this lab you will continue to explore the world of decoders (data receivers). You will also study a device called an encoder and you will see how it is a reverse decoder. Encoders don't receive binary data; they generate binary data.

"1 of X" Decoder

Decoders are **data receivers**. Decoders receive data and generate an output response based on the data received. The decoders used in this lab are called "1 of X" decoders. Refer to Figure 7-1.

A "1 of 4" decoder receives a binary number and asserts one of its outputs. The asserted output is the one that corresponds to the binary number received. For example, if the decoder receives the binary number 2, then it asserts output 2. The decoder shown in Figure 7-1 has

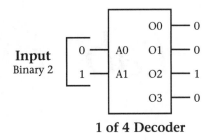

A1	A0	O0	O1	O2	O3
0	0	1	0	0	0
0	1	0	1	0	0
1	0	0	0	1	0
1	1	0	0	0	1

Input	**Output**
2-bit binary number	1 output is active for each binary number received.

Figure 7-1 A "1 of 4" Decoder

active high outputs and can also be called a "2 line to 4 line" decoder. Some decoders have active low outputs. Refer to Figure 7-2.

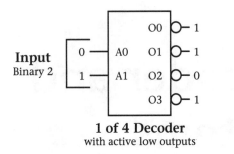

A1	A0	O0	O1	O2	O3
0	0	0	1	1	1
0	1	1	0	1	1
1	0	1	1	0	1
1	1	1	1	1	0

Input	**Output**
2-bit binary number	1 output is active for each binary number received.

Figure 7-2 A "1 of 4" Decoder with Active Low Outputs

Active low outputs are asserted with a 0. Some decoders have an additional control input called ENABLE. The enable input is like a master override. It must be asserted or else all decoder outputs will remain inactive. Refer to Figure 7-3. The decoder of Figure 7-3 works normally if the ENABLE input is connected to 0. If the active low ENABLE input is connected to 1, then all decoder outputs are 1. No outputs are asserted and the inputs A1 and A0 are immaterial. The Xs represent an immaterial (or don't care) condition. Figure 7-4 shows another "1 of 4" decoder. It has active high outputs and an active high enable.

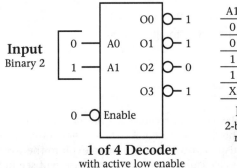

A1	A0	E	O0	O1	O2	O3
0	0	0	0	1	1	1
0	1	0	1	0	1	1
1	0	0	1	1	0	1
1	1	0	1	1	1	0
X	X	1	1	1	1	1

Input	**Output**
2-bit binary number	1 output is active for each binary number received.

Figure 7-3 A "1 of 4" Decoder with Active Low Enable

Some "1 of X" decoders can receive larger binary numbers. They also can have more than one enable input. Figure 7-5 shows a "1 of 8" decoder with three enable inputs. A "1 of 8" decoder also can be called a "3 line to 8" line decoder.

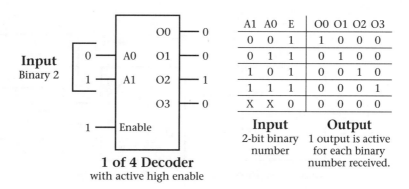

Figure 7-4 A "1 of 4" Decoder with Active High Enable and Active High Outputs

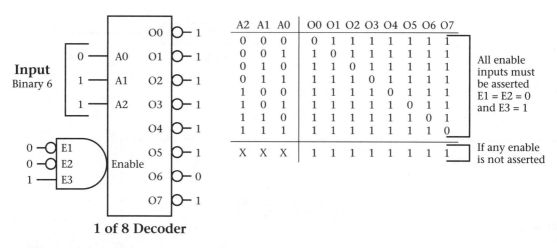

Figure 7-5 A "1 of 8" Decoder with Three Enable Inputs

"1 of X" Decoder Application: Camera Scanner

"1 of X" decoders receive binary data and activate one out of a group of outputs. Figure 7-6 shows how a "1 of 4" decoder can be used in a camera scanner application. A security system has four cameras mounted in four different rooms of a building. At night the security system is used to scan images from each camera onto videotape. Each camera will be enabled for 10 seconds.

VHDL Decoder

The VHDL code for the "1 of 4" decoder with an active high enable (Figure 7-4) is shown in Figure 7-7. The architecture uses a vector statement **SIGNAL input: STD_LOGIC_VECTOR (2 DOWNTO 0);** to define a storage array. The storage array is called **input**, and (2 DOWNTO 0) implies it stores 3 bits of data. The statements **input (2) <= E;**, **input (1) <= A1;**, and **input (0) <= A0;** assign the input variables to the storage array.

The architecture uses a vector statement **SIGNAL output: STD_LOGIC_VECTOR (3 DOWNTO 0);** to define a storage array. The storage array is called **output** and (3 DOWNTO 0) implies it stores 4 bits of data. The statements **O3 <= output(3);**, **O2 <= output(2);**, **O1 <= output(1);**, and **O0 <= output(0);** assign the output array to the output variables O3, O2, O1, and O0.

The "Selected Signal Assignment" statements **WITH input SELECT** will evaluate the input array (E, A1, A0) and assign the output array to "O3," "O2," "O1," and "O0." For example, the statement **output <= "0001" WHEN "100"**, assigns O0 = 1 when E is asserted (active high), and the 2-bit binary number at A1 and A0 is 0. This is the correct response for a decoder. The remaining "Selected Signal Assignment" statements correctly evaluate the remaining decoder input conditions and assign the output responses.

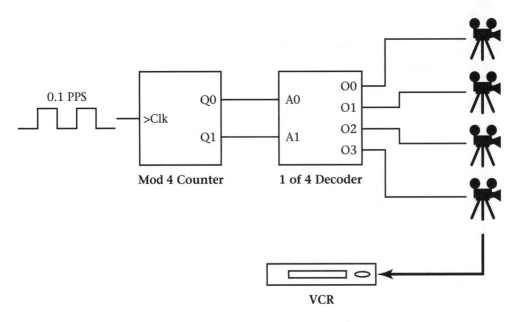

Figure 7-6 A "1 of 4" Decoder Used as a Camera Scanner

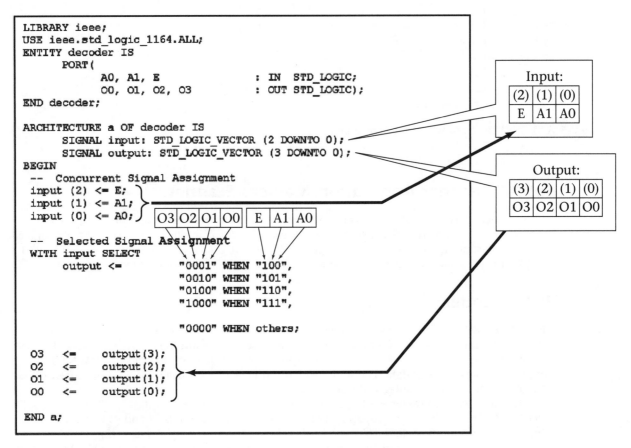

Figure 7-7 VHDL Code for a "1 of 4" Decoder

"1 of X" Encoder

Encoders are **data generators**. Encoders generate binary data based on the assertion of one of its inputs. Refer to the "1 of X" encoder in Figure 7-8. A "1 of 8" encoder generates a binary number that corresponds to the number of the input that is asserted. For example, the encoder in Figure 7-8 generates the binary number 3 because input A3 is asserted. An encoder is a reverse decoder. The encoder shown in Figure 7-8 has active high inputs. Some encoders have active low inputs. Refer to Figure 7-9. An encoder with active low inputs generates binary data at the output based on which input is asserted with a 0. Some encoders eliminate the A0 input. Refer to Figure 7-10.

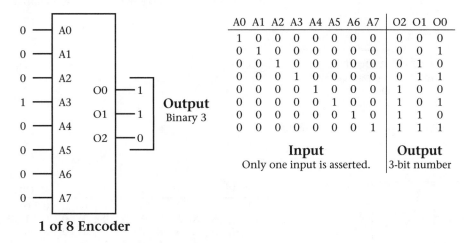

A0	A1	A2	A3	A4	A5	A6	A7	O2	O1	O0
1	0	0	0	0	0	0	0	0	0	0
0	1	0	0	0	0	0	0	0	0	1
0	0	1	0	0	0	0	0	0	1	0
0	0	0	1	0	0	0	0	0	1	1
0	0	0	0	1	0	0	0	1	0	0
0	0	0	0	0	1	0	0	1	0	1
0	0	0	0	0	0	1	0	1	1	0
0	0	0	0	0	0	0	1	1	1	1

Input	**Output**
Only one input is asserted.	3-bit number

1 of 8 Encoder

Figure 7-8 A "1 of 8" Encoder

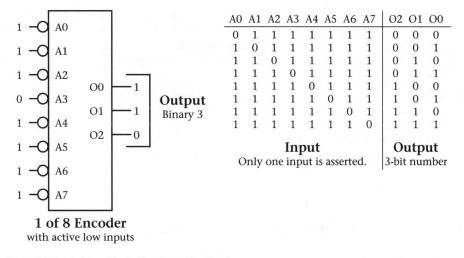

A0	A1	A2	A3	A4	A5	A6	A7	O2	O1	O0
0	1	1	1	1	1	1	1	0	0	0
1	0	1	1	1	1	1	1	0	0	1
1	1	0	1	1	1	1	1	0	1	0
1	1	1	0	1	1	1	1	0	1	1
1	1	1	1	0	1	1	1	1	0	0
1	1	1	1	1	0	1	1	1	0	1
1	1	1	1	1	1	0	1	1	1	0
1	1	1	1	1	1	1	0	1	1	1

Input	**Output**
Only one input is asserted.	3-bit number

1 of 8 Encoder
with active low inputs

Figure 7-9 A "1 of 8" Encoder with Active Low Inputs

The encoder in Figure 7-10 will output the code for "0" if all inputs "A1" through "A7" are not asserted. The encoder has a default value to "0" when all other inputs are not asserted. This eliminates the need for an additional input ("A0") while giving the system the ability to generate the code for "0." Some encoders are called priority encoders. Priority encoders prioritize the order of the data generated at the output. Refer to Figure 7-11.

If more than one input is asserted, the priority encoder shown in Figure 7-11 generates the binary number for the largest asserted input.

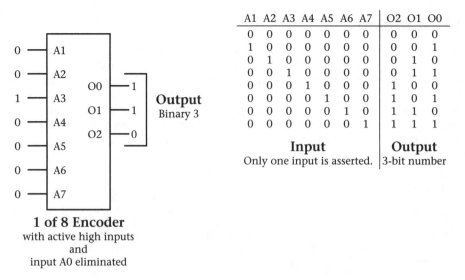

A1	A2	A3	A4	A5	A6	A7	O2	O1	O0
0	0	0	0	0	0	0	0	0	0
1	0	0	0	0	0	0	0	0	1
0	1	0	0	0	0	0	0	1	0
0	0	1	0	0	0	0	0	1	1
0	0	0	1	0	0	0	1	0	0
0	0	0	0	1	0	0	1	0	1
0	0	0	0	0	1	0	1	1	0
0	0	0	0	0	0	1	1	1	1

Input	Output
Only one input is asserted.	3-bit number

1 of 8 Encoder
with active high inputs
and
input A0 eliminated

Figure 7-10 A "1 of 8" Encoder without A0 Input

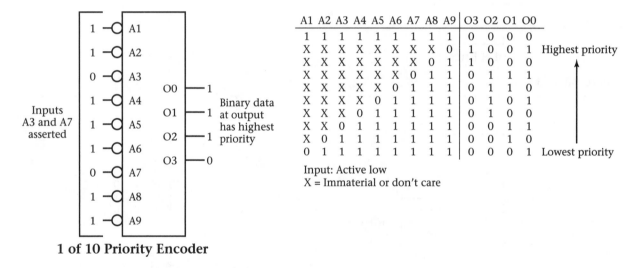

A1	A2	A3	A4	A5	A6	A7	A8	A9	O3	O2	O1	O0	
1	1	1	1	1	1	1	1	1	0	0	0	0	
X	X	X	X	X	X	X	X	0	1	0	0	1	Highest priority
X	X	X	X	X	X	X	0	1	1	0	0	0	
X	X	X	X	X	X	0	1	1	0	1	1	1	
X	X	X	X	X	0	1	1	1	0	1	1	0	
X	X	X	X	0	1	1	1	1	0	1	0	1	
X	X	X	0	1	1	1	1	1	0	1	0	0	
X	X	0	1	1	1	1	1	1	0	0	1	1	
X	0	1	1	1	1	1	1	1	0	0	1	0	
0	1	1	1	1	1	1	1	1	0	0	0	1	Lowest priority

Input: Active low
X = Immaterial or don't care

1 of 10 Priority Encoder

Figure 7-11 A "1 of 10" Priority Encoder

"1 of X" Encoder Application: SPST Keypad

Encoders can be used with SPST (single pole single throw) keypads to convert a pressed button into a binary number. Refer to Figure 7-12. The switch that is pressed will assert an encoder input and will generate the 4-bit switch number at the encoder output. The switches that are not pressed have open contacts and pull-up resistors that will make the encoder inputs 1 (not asserted).

VHDL Encoder

The VHDL code for the 1 of 10 priority encoder (see Figure 7-11) is show in Figure 7-13. The architecture uses a vector statement **SIGNAL output: STD_LOGIC_VECTOR (3 DOWNTO 0);** to define a storage array. The storage array is called **output**, and **(3 DOWNTO 0)** implies it stores 4 bits of data. The statements **O3 <= output(3);, O2 <= output(2);, O1 <= output(1);,** and **O0 <= output(0);** assign the output array to the output variables O3, O2, O1, and O0.

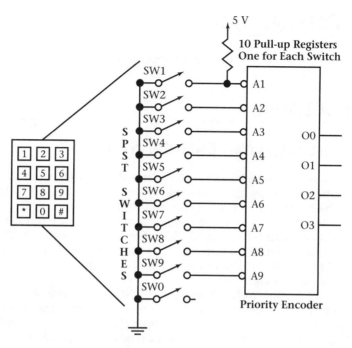

Figure 7-12 A "1 of 10" Priority Encoder Connected to an SPST Keypad

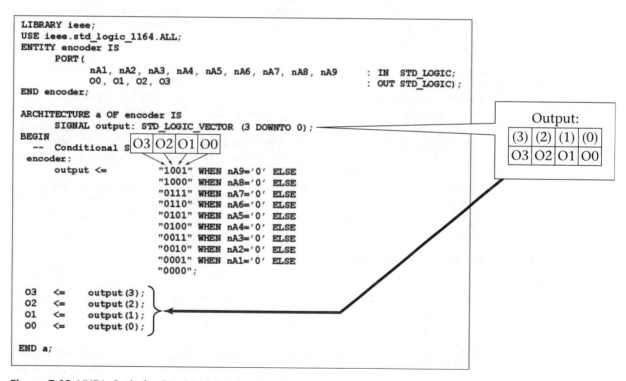

Figure 7-13 VHDL Code for "1 of 10" Priority Encoder

The VHDL **conditional signal assignment** statement **output <= "1001" WHEN nA9='0' ELSE** creates the priority scheme required by the encoder. **nA9** has the highest priority and sets the output to "9" when "nA9" is asserted ("nA9" = 0). The **ELSE** statements of the lower priority inputs ("nA8," "nA7," ... "nA1") are only evaluated if a higher priority input is *not* asserted. If no inputs are asserted, then "0" is output. This is the correct response for a priority encoder.

POWERPOINT PRESENTATION

Use PowerPoint to view the Lab 07 slide presentation.

LAB WORK PROCEDURE

Build and Test A Keypad Encoder System

An SPST keypad will be connected to an encoder, a data register, and a display decoder. The system will store and display the number of the key that is pressed. This lab can be completed without an SPST keypad. MAX switches can be substituted for the SPST keypad.

1. The debounce files from Lab 3 are required to complete this lab. A VHDL 1 of 10 priority encoder file on the CD-ROM needs to be placed onto the disk that contains the Lab 3 debounce files. There is a folder (directory) on your disk named **Labs** that was used to store the debounce files. Copy the file **encoder.vhd** from the CD-ROM to the folder **Labs** on your disk.

   As you have seen in Lab 2, to create a VHDL design you must create a text file to enter the VHDL code. The text file is converted to a symbol that can be displayed in the Graphic Editor. The Graphic Editor is used to connect input and output symbols. Effectively, your VHDL code creates a new symbol in the Altera symbol library. The VHDL text file representing the encoder has been copied from the CD-ROM. You need only create a symbol for the VHDL encoder.

2. Start the **MAX+PLUS II** program. Open the file **encoder.vhd**. Use the section "Lab Work Procedure: VHDL Vending Machine" in Lab 2 as a guide. Review steps 1 through 5. These steps are not needed because the VHDL file was copied from the CD-ROM. You need to open the VHDL file and complete step 6. Step 6 creates the symbol. Ignore any warning message.

   The text file that represents the VHDL encoder has been converted into a symbol that can be displayed in the Graphic Editor.

3. You will draw a diagram for the keypad encoder system. Use the Lab Work Procedure in Lab 1 as a guide, and complete steps 1 through 12. Use the project name **Lab7.gdf**. The 4-bit data register symbol in the "mega_lpm" library is called **lpm_DFF**. The OR gate symbol in the "prim" library is called **OR6**. The display decoder in the "mf" library is called **7447** (or you can use the display decoder from Lab 6). The Graphic Editor file for the keypad encoder should resemble Figure 7-14.

4. The UP board can now be programmed. Lab 1 describes all the steps that are required to program a UP board with the keypad encoder system. Use the Lab Work Procedure in Lab 1 as a guide, and complete steps 13 through 17. Ignore any compiler warning and information messages.

MAX and FLEX IC Considerations

MAX IC Design:

- Connect **Switch0** through **Switch9** to an **SPST keypad** (or **MAX** switches if you don't have a keypad).
- Assign input **Oscillator** to pin number **83**. Refer to Lab 1, step 13_{FLEX} for assistance in assigning an input symbol to a pin number.
- Assign pin numbers to output symbols **SegA**, **SegB**, ... **SegG** to connect the **MAX** digital display to the display decoder. Don't forget to turn off all of the other unused LEDs in order to block the MAX+PLUS II software from randomly routing logic signals through these pins. Refer to "Lab 6: Lab Work Procedure: Add a Display Decoder to the Random Number Generator System" for assistance.
- You are ready to test the operation of the keypad encoder system.

FLEX IC Design:

- Assign **Switch0** through **Switch9** to the **FLEX Expansion Header**. Connect **Switch0** through **Switch9** to an SPST keypad (or connect wires over to the MAX switches if you don't have a keypad). Appendix C describes the "FLEX Expansion Header" in more detail.

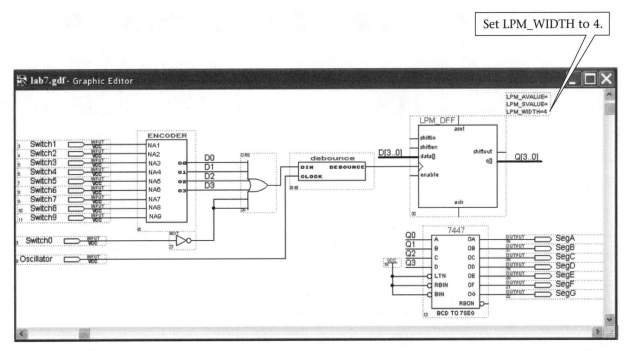

Figure 7-14 Altera "Graphic Editor": Keypad Encoder System

- Assign input **Oscillator** to pin number **91** to connect to the 25,175,000 PPS UP board oscillator.
- Assign pin numbers to output symbols **SegA**, **SegB**, ... **SegG** to connect the **FLEX** digital display to the display decoder. Don't forget to turn off all of the other unused LEDs.
- You are ready to test the operation of the keypad encoder system.

LAB EXERCISES AND QUESTIONS

This section contains lab exercises that can be performed on the UP board and questions that can be answered at home. Ask your instructor which exercises to perform and which questions to answer.

Lab Exercises

Exercise 1: Test the keypad encoder system.
Test the operation of the keypad system and answer the following questions:
   a. Explain how switch *bounce* affects the *encoder*. How does the output of the encoder respond to a bouncing input switch?
   b. What is the function of the *OR* gate? When does it output a "0"? When does it output a "1"?
   c. How does the system store and display "key 0" when it is not even connected to the encoder?
   d. Explain how the *keypad*, the *encoder*, the *OR gate*, the *debounce* system, the *data register*, and the *display decoder* all work together to store and display a key number.
   e. Use the UP board to change the system, observe its operation, and answer a question. Bypass the debounce system and recompile and reprogram (or reconfigure) the Altera IC. Refer to Figure 7-15.
   How well does the system work if the debounce system is bypassed and the OR gate is directly connected to the clock of the data register? Try pressing several keys. You may need to pick one key and quickly and repeatedly tap it to see any effect of switch bounce. Be sure to reconnect the debounce system after this test is complete.

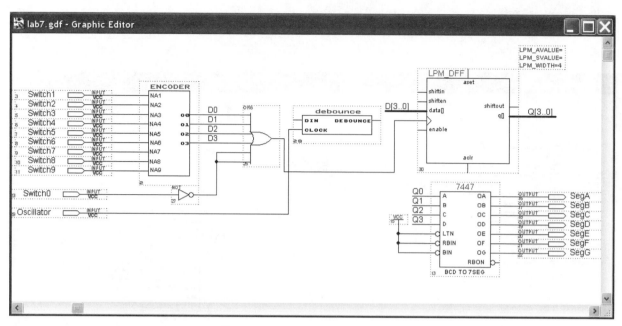

Figure 7-15 Altera "Graphic Editor": Keypad Encoder with Debounce System Bypassed.

Exercise 2: Test the keypad encoder system without memory.

Use the UP board to change the system, observe its operation, and answer questions. The display decoder input (D, C, B, and A) will be directly connected to the encoder output (D0, D1, D2, and D3), bypassing the OR gate and data register. Change the system diagram to effectively bypass the OR gate and the data register and recompile and reprogram (or reconfigure) the Altera CPLD. Refer to Figure 7-16.

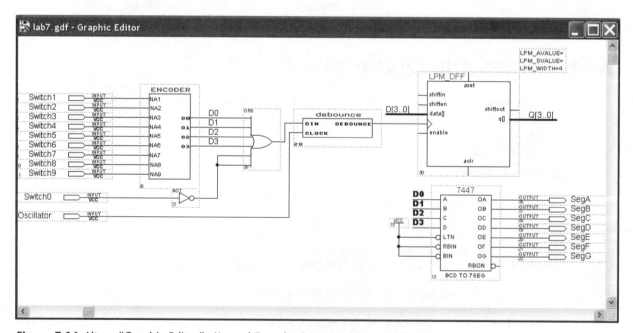

Figure 7-16 Altera "Graphic Editor": Keypad Encoder System without Memory

    a. Explain how the system works.
    b. Try pressing and holding down two keys at the same time. Explain how the system responds to this condition.
    c. Explain the effect that switch bounce has on the system.

Exercise 3: Design a two-key storage system

Use the UP board to change the system and observe its operation. Add a second display decoder and a second data register to create a system capable of storing and displaying two separate key numbers. The first key is displayed on the first digital display and the second key is displayed on the second digital display. You will need to use some creative thinking to control the clock inputs of the two data registers. You do not want one key to be displayed twice. Be sure to use the ripple blanking feature to turn off leading zeros. For example, if the numbers 07 are stored, they will be displayed as the number 7 without a leading 0.

Exercise 4: Design and test the camera scanner system.

Design and test the operation of a camera scanner system. The system is described in the Introductory Information section of this lab. The CD-ROM contains a VHDL decoder file: **decoder.vhd**. Use *LEDs* to represent *cameras*. Use a scan rate of approximately "1 second" instead of "10 seconds."

Exercise 5: Use the Altera simulator to test the camera scanner system.

(This exercise can be done at home.)

Use the MAX+PLUS II "Graphic Editor" to make changes to the "camera scanner" system. Make sure inputs **Decoder_A0** and **Decoder_A1** are connected to the **LPM_COUNTER** outputs **Q0** and **Q1**. The change will allow you to use the simulator to test its operation. Use Appendix F as a guide to run a simulation for the camera scanner system you just created. The changes may give warnings when you use "Save, Compile & Simulate." These warnings can be ignored. The "Waveform Editor" window for the simulation should resemble Figure 7-17.

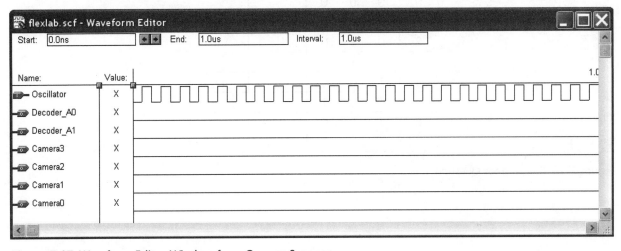

Figure 7-17 Waveform Editor Window for a Camera Scanner

Set the **End time** to 1.0us. The "Oscillator" uses an "Overwrite clock period" = 40.0ns. If the "Clock Period" entry box is grayed out (not selectable), then you should exit the "Overwrite Clock" window, select the Options menu, and remove the check mark beside the menu item "Snap to Grid." Return to the "Overwrite Clock" window and you will be able to set the clock period.

a. Run the simulation and explain the output response you observe at the "camera" outputs.

b. Create two node groups. Group the "Decoder_A" inputs into a single node. Remember to reorder the waveforms with the most significant waveform at the top. Group the "Camera" outputs. Run the simulation again and explain the output response you observe at the "camera" outputs.

Lab Questions

1. Analyze the system shown in Figure 7-18 and fill in the table.

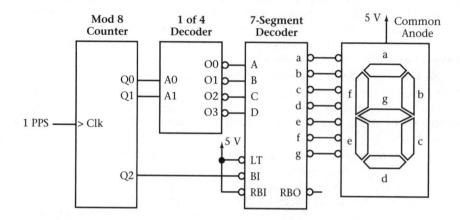

Clk	Q2	Q1	Q0	DCBA	Display
0	0	0	0		
1	0	0	1		
2	0	1	0		
3	0	1	1		
4	1	0	0		
5	1	0	1		
6	1	1	0		
7	1	1	1		

Figure 7-18 Diagram for Question 1

2. Analyze the system shown in Figure 7-19 and fill in the table.

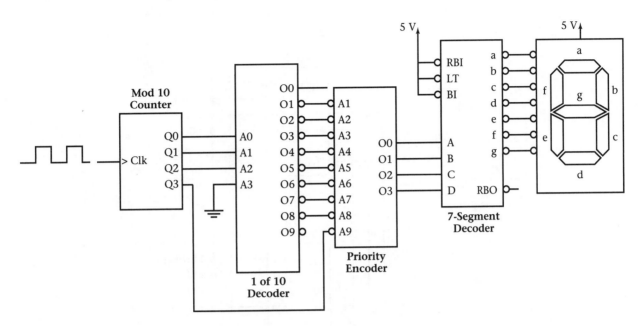

Clk	Q3	Q2	Q1	Q0	Decoder A3 ... A0	Display
0	0	0	0	0		
1	0	0	0	1		
2	0	0	1	0		
3	0	0	1	1		
4	0	1	0	0		
5	0	1	0	1		
6	0	1	1	0		
7	0	1	1	1		
8	1	0	0	0		
9	1	0	0	1		

Figure 7-19 Diagram for Question 2

3. Analyze the system shown in Figure 7-20 and fill in the table.

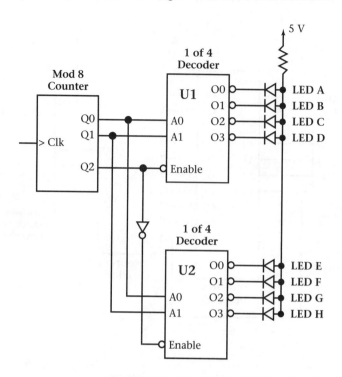

Q2 Q1 Q0	U1				U2				Lit LED
	O0	O1	O2	O3	O0	O1	O2	O3	
0 0 0									
0 0 1									
0 1 0									
0 1 1									
1 0 0									
1 0 1									
1 1 0									
1 1 1									

Figure 7-20 Diagram for Question 3

4. The system in Figure 7-21 shows how a decoder can be used in the generation of control signals. Assume the counter begins at 0000. Sketch the waveforms for the counter along with O0 and O1 of the decoder.

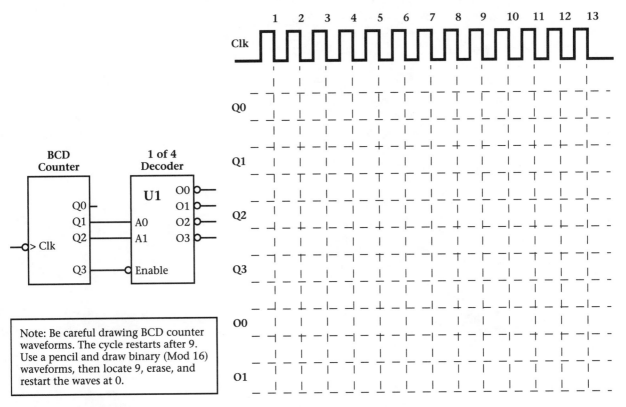

Note: Be careful drawing BCD counter waveforms. The cycle restarts after 9. Use a pencil and draw binary (Mod 16) waveforms, then locate 9, erase, and restart the waves at 0.

Figure 7-21 Diagram for Question 4

5. Write the VHDL code for a "1 of 8" decoder with an active high enable and active low outputs. Use the VHDL "1 of 4" decoder shown in Figure 7-7 as a guide.

CONCLUSIONS

- Decoders are *data receivers*. A decoder receives a binary number and asserts one of its outputs.
- If a decoder "ENABLE" input is not asserted, then all decoder outputs will remain inactive.
- A camera scanner system can be created from a decoder.
- Encoders are *data generators*. An encoder generates a binary number that corresponds to the number of the input that is asserted.
- A priority encoder generates the binary data for the largest asserted input.
- An encoder can be used to encode a keypad.

Lab 8

Multiplexer and Demultiplexer Systems

Equipment   Altera UP board
Book CD-ROM
Spool of 24 AWG wire
Wire cutters
A blank floppy disk

Objectives   Upon completion of this lab, you should be able to:

- Operate a multiplexer and a demultiplexer.
- Build a security system using a multiplexer and a demultiplexer.
- Understand *time division multiplexing*.
- Use a single multiplexer to design an entire logic gate system.

INTRODUCTORY INFORMATION

In previous labs, decoders and encoders were used to receive and generate data. These devices were part of a family of digital devices called **data handlers**. In this lab you will add two more devices to this family of data handlers. They are called the **multiplexer** and the **demultiplexer**. The *multiplexer* is called a **data selector**. Its operation is analogous to the channel selecting system found inside a television set. The *demultiplexer* is a **data distributor**. Its operation is analogous to a job performed by the dealer when a card game is played. The dealer distributes the cards to each player.

The Multiplexer (MUX)

A multiplexer is a data selector. It is also called a **mux**. A mux receives binary data from many sources and selects one of the sources to transmit its data to the output. The sources of binary data are called channels. Refer to Figure 8-1. A MUX is like an electronic selector switch. The 2-bit binary number on the **select control inputs** selects the input channel that transmits its data to output Z. Refer to Figure 8-2.

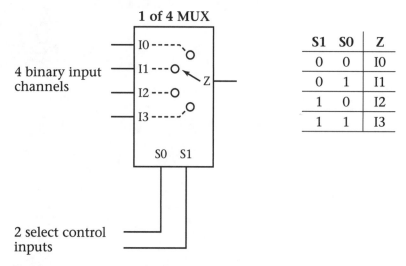

S1	S0	Z
0	0	I0
0	1	I1
1	0	I2
1	1	I3

Figure 8-1 A "1 of 4" MUX

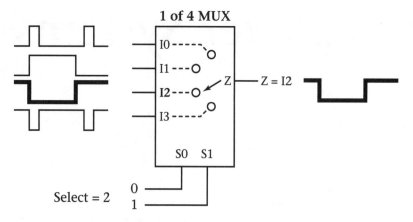

Figure 8-2 A "1 of 4" MUX Selecting Channel 2

The MUX in Figure 8-2 has the number 2 on the "select" inputs. "Z" will transmit the binary information on channel 2 (Z = I2). The other three channels ("I0," "I1," and "I3") are ignored. To transmit the other channels, you must change the 2-bit number at "S1" and "S0." The **1 of 4** MUX is also called a **4 line to 1 line MUX** or a **4 to 1** MUX. Some MUXs have an additional control input called **ENABLE**. The enable input is like a master override. If it is not asserted, then output "Z" will remain inactive. Refer to Figure 8-3.

Figure 8-4 shows a "1 of 8" MUX with an active low enable and an active high and active low output. The active low output inverts the data of the selected channel. Its use is optional.

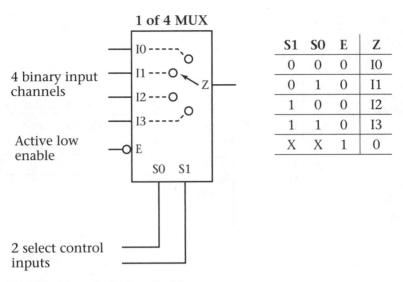

Figure 8-3 A "1 of 4" MUX with an Active Low Enable

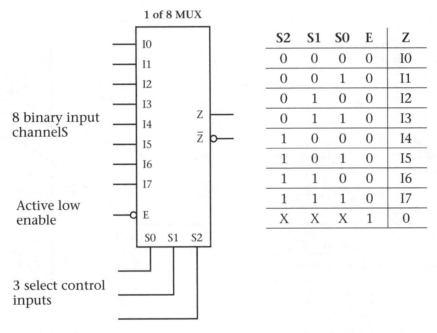

Figure 8-4 A "1 of 8" MUX with an Active Low Enable and an Active High and Active Low Output

Figure 8-5 shows a "Quad 1 of 2" MUX with an active low enable. Each channel is made up of a 4-bit number. Only one select input is needed because this line chooses to send a 4-bit number from I0 to Z or from I1 to Z.

A VHDL MUX

The VHDL code for the "1 of 4" MUX with an active low enable (see Figure 8-3) is shown in Figure 8-6. The architecture uses a vector statement **SIGNAL input: STD_LOGIC_VECTOR (2 DOWNTO 0);** to define a storage array. The storage array is called **input**, and **(2 DOWNTO 0)** implies it stores 3 bits of data. The statements **input (2) <= nE;, input (1) <= S1;,** and **input (0) <= S0;** assign the input variables to the storage array.

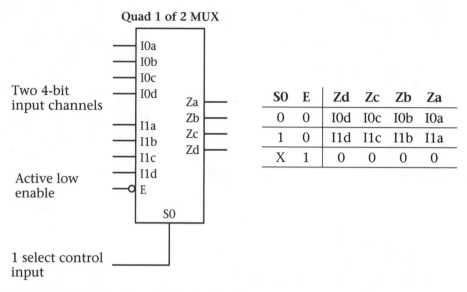

Figure 8-5 A Quad "1 of 2" MUX with an Active Low Enable

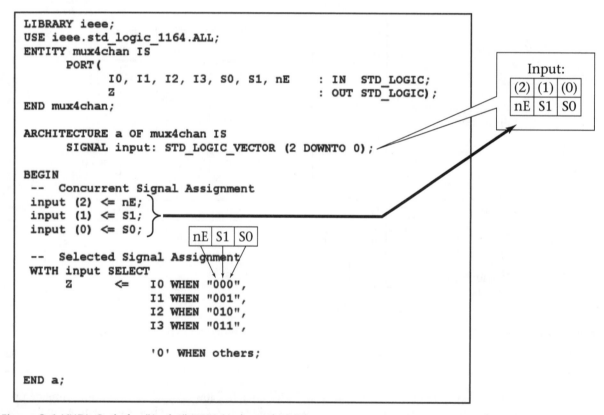

Figure 8-6 VHDL Code for "1 of 4" MUX (4-channel MUX)

The **Selected Signal Assignment** statements **WITH input SELECT** will evaluate the input array (**nE, S1, S0**) and assign "I0," "I1," I2, or "I3" to the output "Z." For example, the statement **Z <= I0 WHEN "000"** assigns **I0** to Z when **nE** is asserted (active low) and the 2-bit binary number at **S1** and S0 is 0. This is the correct response for a multiplexer. The remaining "Selected Signal Assignment" statements evaluate the remaining multiplexer input conditions and correctly assign "Z."

The Demultiplexer (DEMUX)

A demultiplexer is a data distributor. It is also called a **DEMUX**. A DEMUX receives binary data from one source and transmits its data to the selected output. It is a reverse multiplexer. Refer to Figure 8-7.

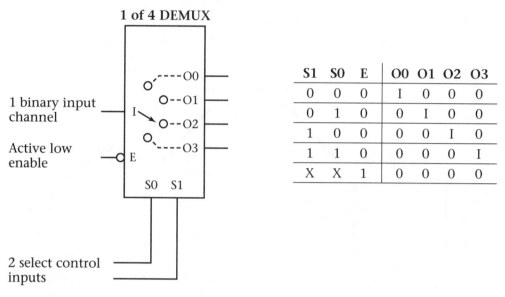

S1	S0	E	O0	O1	O2	O3
0	0	0	I	0	0	0
0	1	0	0	I	0	0
1	0	0	0	0	I	0
1	1	0	0	0	0	I
X	X	1	0	0	0	0

Figure 8-7 A "1 of 4" DEMUX

Figure 8-8 shows a "1 of 8" DEMUX with an active low enable. A DEMUX can be constructed from a decoder. This was once an important feature that allowed older IC technology manufacturers to fabricate one device and use it two different ways. The device was called a decoder/demultiplexer. Refer to Figure 8-9.

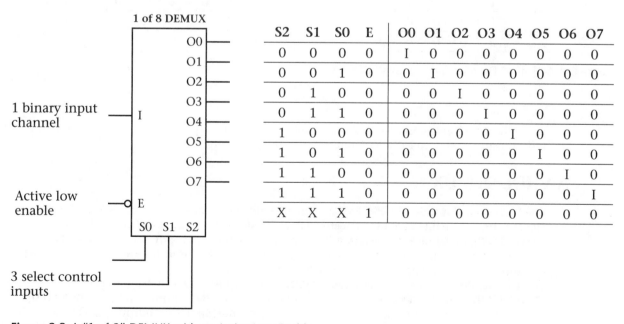

S2	S1	S0	E	O0	O1	O2	O3	O4	O5	O6	O7
0	0	0	0	I	0	0	0	0	0	0	0
0	0	1	0	0	I	0	0	0	0	0	0
0	1	0	0	0	0	I	0	0	0	0	0
0	1	1	0	0	0	0	I	0	0	0	0
1	0	0	0	0	0	0	0	I	0	0	0
1	0	1	0	0	0	0	0	0	I	0	0
1	1	0	0	0	0	0	0	0	0	I	0
1	1	1	0	0	0	0	0	0	0	0	I
X	X	X	1	0	0	0	0	0	0	0	0

Figure 8-8 A "1 of 8" DEMUX with an Active Low Enable

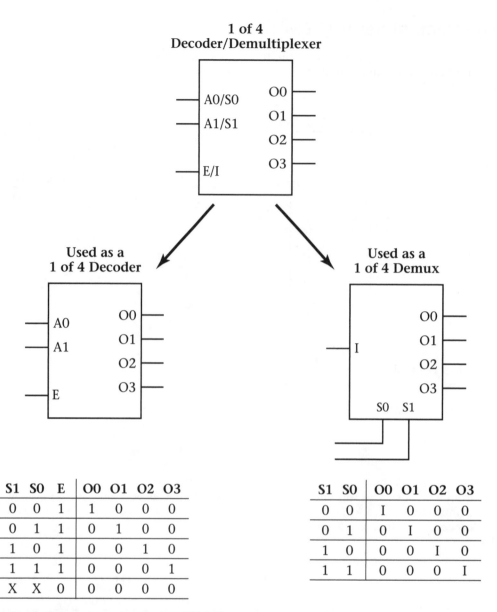

Figure 8-9 A "1 of 4" Decoder and a "1 of 4" DEMUX

With the advent of CPLDs, this dual functionality is less important, but understanding the linked history that decoders and demultiplexers share is useful.

A VHDL DEMUX

The VHDL code for the "1 of 4" DEMUX with an active low enable (see Figure 8-7) is shown in Figure 8-10. The architecture uses a vector statement **SIGNAL input: STD_LOGIC_VECTOR (3 DOWNTO 0);** to define a storage array. The storage array is called **input**, and (3 DOWNTO 0) implies it stores 4 bits of data. The statements **input (3) <= nE;, input (2) <= I;, input (1) <= S1;,** and **input (0) <= S0;** assign the input variables to the storage array.

The architecture uses a vector statement **SIGNAL output: STD_LOGIC_VECTOR (3 DOWNTO 0);** to define a storage array. The storage array is called **output**, and (3 DOWNTO 0) implies it stores 4 bits of data. The statements **O3 <= output(3);, O2 <= output(2);, O1 <= output(1);,** and **O0 <= output(0);** assign the **output** array to the **output** variables O3, O2, O1, and O0.

The "Selected Signal Assignment" statements **WITH input SELECT** will evaluate the input array ("nE," "I," "S1," "S0") and assign the output array to "O3," "O2," "O1" and "O0." For

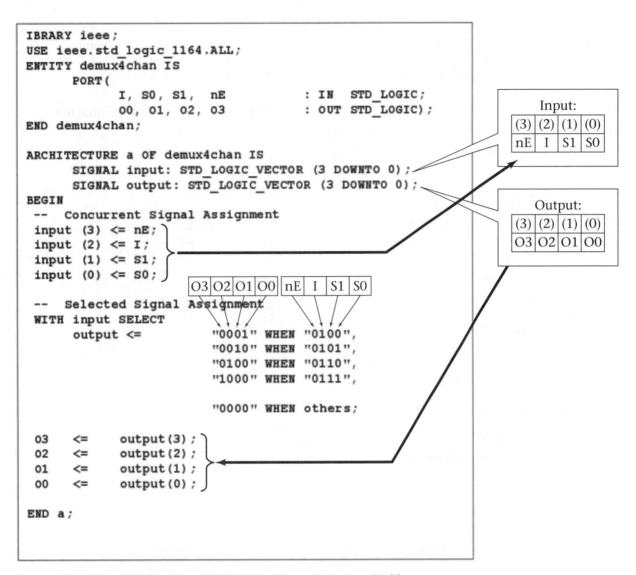

```
IBRARY ieee;
USE ieee.std_logic_1164.ALL;
ENTITY demux4chan IS
      PORT(
              I, S0, S1,  nE           : IN  STD_LOGIC;
              O0, O1, O2, O3           : OUT STD_LOGIC);
END demux4chan;

ARCHITECTURE a OF demux4chan IS
      SIGNAL input: STD_LOGIC_VECTOR (3 DOWNTO 0);
      SIGNAL output: STD_LOGIC_VECTOR (3 DOWNTO 0);
BEGIN
--  Concurrent Signal Assignment
input (3) <= nE;
input (2) <= I;
input (1) <= S1;
input (0) <= S0;

--  Selected Signal Assignment
WITH input SELECT
      output <=           "0001" WHEN "0100",
                          "0010" WHEN "0101",
                          "0100" WHEN "0110",
                          "1000" WHEN "0111",

                          "0000" WHEN others;

O3    <=    output(3);
O2    <=    output(2);
O1    <=    output(1);
O0    <=    output(0);

END a;
```

Input:

(3)	(2)	(1)	(0)
nE	I	S1	S0

Output:

(3)	(2)	(1)	(0)
O3	O2	O1	O0

O3	O2	O1	O0	nE	I	S1	S0

Figure 8-10 VHDL Code for a "1 of 4" DEMUX with an Active Low Enable

example, the statement **output <=** "0001" WHEN "0100", assigns O0 = 1 when **nE** is asserted (active low), **I** = "1," and the 2-bit binary number at **S1** and **S0** is **0**. This is the correct response for a demultiplexer. The remaining "Selected Signal Assignment" statements correctly evaluate the remaining demultiplexer input conditions and assign the output responses. The statement "0000" WHEN others; assigns 0 to all outputs of demultiplexer, including the selected output channel, regardless of the input conditions. This is the correct response for a demultiplexer.

A MUX/DEMUX Security System

A security system can be constructed using a MUX and DEMUX. Refer to Figure 8-11. The security system monitors the open/closed status of four doors. A detailed description of the operation of this system will follow. In short, when a door is opened, the door sensor sends a binary 1 to the MUX. The MUX, the DEMUX, and the counters work together to send the binary 1 to the LED. The LED will light to indicate the door is opened. A MUX/DEMUX configuration minimizes the amount of wiring from the door sensors to the door LEDs. This four-door security system sends the door status information to an LED panel using four wires. A 16-door security would require a larger MUX (1 of 16), a larger DEMUX (1 of 16), and a four-stage counter (Mod 16). The 16-door security system would require only two extra wires because of the larger counter. For each additional counter output, the system doubles the number of doors it can monitor. This is the main benefit of a MUX/DEMUX security system. It should be noted that there are other techniques that can be used to reduce the wiring even more, for example, if the MUX and DEMUX each had their own counters synchronized with a common clock source. The number of doors monitored by the system would not increase the number of wires.

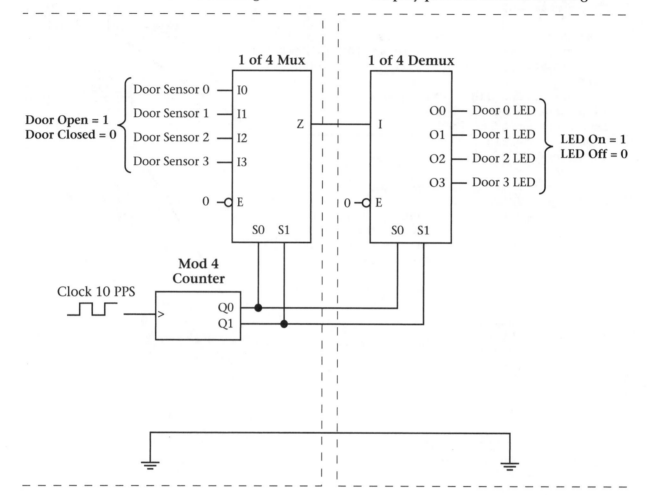

Q1	Q0	Z
0	0	I0
0	1	I1
1	0	I2
1	1	I3

Q1	Q0	O0	O1	O2	O3
0	0	I0	0	0	0
0	1	0	I1	0	0
1	0	0	0	I2	0
1	1	0	0	0	I3

Figure 8-11 MUX/DEMUX Security System

Security System Detailed Operation

The 10 PPS clock signal causes each count state to last one-tenth of a second. Assume the counters start at 0. For one-tenth of a second the MUX is sending the data from **Door Sensor 0** to DEMUX. The DEMUX is sending **Door Sensor 0** data to **Door 0 LED**. If **Door 0** is open then the LED is on for one-tenth of a second. Meanwhile the door activity for doors 1, 2, 3 are not being monitored and LEDs 1, 2, and 3 are off.

When the counter increments to "1," then **Door Sensor 1** data is being sent to **Door 1 LED** for one-tenth of a second. The count cycle continues with doors 2 and 3 sending data to LEDs 2 and 3. The counter scans each door for one-tenth of a second and then the entire cycle is repeated over and over. The clock signal controls the scan rate. At 10 PPS an open door will cause an LED to blink. The LED is on for 1/10th of a second and is off for three-tenths of a second. A blinking LED is desirable because it attracts attention. Decreasing the scan rate even more would result in a slower blinking LED. It can be argued that the slower blink rate could make the blinking LED even more noticeable; however, the trade-off is security. If the pulse rate is too slow, then the time taken to scan all the doors could be long and could possibly result in a missed intrusion. Increasing the scan rate would result in a more secure system; however, a high pulse per second rate can eventually result in the loss of the blinking LED.

Time Division Multiplexing

It is worth noting that a MUX/DEMUX security system is an example of a **TDM** system. **TDM** is short for **time division multiplexing**. TDM is used by phone companies to maximize the use of their phone lines. Instead of door status being sent through a MUX/DEMUX, think of phone users' voices being sent through the MUX and DEMUX system. Sharing the communication channel with many users allows more conversations to be sent over a single phone line. This feature allows a phone company to use a single communication line efficiently. What limits adding more users to a phone system is the scan rate speed. The MUX and DEMUX must be synchronized and scanned fast enough in order not to degrade the quality of phone conversation.

Using A MUX As A Logic Gate System

A single MUX can replace an entire logic gate system. This was once an important fact that allowed older IC technology designers to replace many single-function logic gate ICs with a single MUX IC. Because you may encounter this design technique, it is useful to understand how it works. Figure 8-12 shows how to use a MUX to implement a digital system.

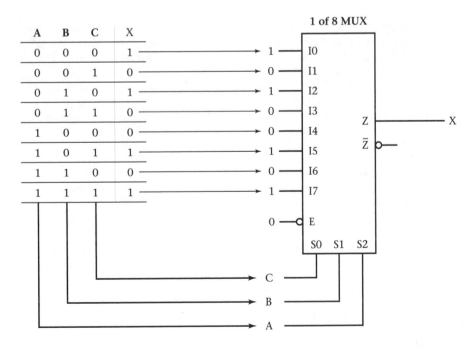

Figure 8-12 Using a Mux as a Logic System

The following is the procedure to design a MUX logic gate system:
1. The MUX channel inputs are connected directly to 1 or 0 as dictated by output X of the truth table.

2. The MUX select inputs become the input of the logic gate system.
3. The MUX output becomes the logic gate output X.

If you study the operation of the system, you will see that the logic gate inputs select a MUX channel. The selected channel sends a constant 1 or 0 to the MUX output. The MUX output matches the truth table response.

POWERPOINT PRESENTATION

Use PowerPoint to view the Lab 08 slide presentation.

LAB WORK PROCEDURE:

Design a MUX/DEMUX Security System

You will use two UP boards to test a MUX/DEMUX security system. The first UP board will be the door interface multiplexer and the second will be the LED panel demultiplexer. Linking the two UP boards requires wiring access to the Altera IC. UP boards are shipped from the factory with wire headers on the MAX IC but not on the FLEX IC. The FLEX IC requires that you solder a flex expansion header socket to the UP board. Appendix C describes the FLEX expansion header in more detail.

Door Interface Multiplexer

A four-channel multiplexer will be used to scan the doors.

1. A VHDL four-channel multiplexer file on the CD-ROM needs to be placed onto a blank disk. Begin by creating a folder (directory) on your blank disk named **Labs**. Copy the file **mux4chan.vhd** from the CD-ROM to folder **Labs** on your disk.
    As you have seen in Lab 2, to create a VHDL design you must create a text file to enter the VHDL code. The text file is converted to a symbol that can be displayed in the Graphic Editor. The Graphic Editor is used to connect input and output symbols. Effectively, your VHDL code creates a new symbol in the Altera symbol library. The VHDL text file representing the MUX has been copied from the CD-ROM. You need only create a symbol for the VHDL MUX.
2. Start the **MAX+PLUS II** program. Open the file **mux4chan.vhd**. Use the section "Lab Work Procedure: VHDL Vending Machine" in Lab 2 as a guide. Review steps 1 through 5. These steps are not needed because the VHDL file was copied from the CD-ROM. You need to open the VHDL file and complete step 6. Step 6 creates the symbol. Ignore any warning message.
    The text file that represents your VHDL MUX has been converted into a symbol that can be displayed in the Graphic Editor.
3. You will draw a diagram for the door interface multiplexer system. Use the Lab Work Procedure in Lab 1 as a guide, and complete steps 1 through 12. Use the project name **Lab8**. The counter symbol in the "mega_lpm" library is called **lpm_counter**. The Graphic Editor file for the door interface multiplexer should resemble Figure 8-13.
4. The UP board can now be programmed. Lab 1 describes all the steps that are required to program a UP board with the door interface multiplexer system. Use the Lab Work Procedure in Lab 1 as a guide, and complete steps 13 through 17. Ignore any compiler warning and information messages.

MAX and FLEX IC Considerations

MAX IC Design:
-   Connect input **Door0**, **Door1**, **Door2**, and **Door3** to a UP board **MAX** switches.
-   Assign input **Oscillator** to pin number 83. A PCB trace connects the 25,175,000 PPS oscillator to pin 83 of the MAX IC. Refer to Lab 1, step 13$_{FLEX}$ for assistance in assigning an input symbol to a pin number.
-   *Do not* connect outputs "Mux_out," "Q22," and "Q23" at this time. These outputs will be used to link this system to the LED panel demultiplexer UP board.

Set LPM_WIDTH to 32.

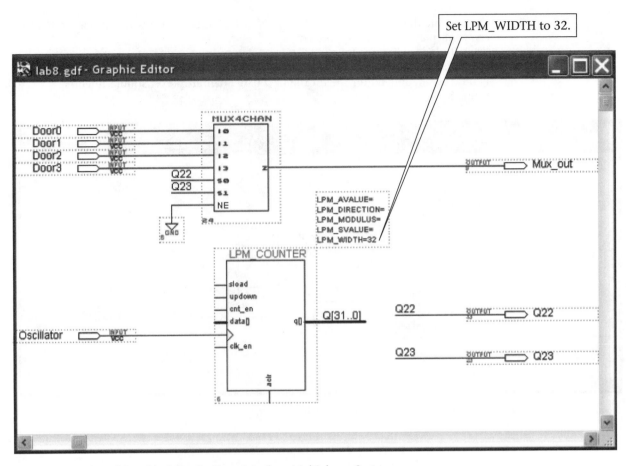

Figure 8-13 Altera "Graphic Editor": Door Interface Multiplexer System

FLEX IC Design:
- Assign input **Door0**, **Door1**, **Door2**, and **Door3** to a UP board **FLEX** switches.
- Assign input **Oscillator** to pin number **91**. A PCB trace connects the 25,175,000 PPS oscillator to pin 91 of the FLEX IC.
- Assign outputs **Mux_out**, **Q22**, and **Q23** to the **FLEX expansion header.** It will allow you to link this system to the LED panel demultiplexer UP board. Appendix C describes the "FLEX expansion header" in more detail. Do not connect wires to these pins at this time.

You are ready to test the operation of the door interface multiplexer system.

LED Panel Demultiplexer

A four-channel demultiplexer will be used to display the door status.
1. A VHDL four-channel demultiplexer file on the CD-ROM needs to be placed onto a blank disk. Begin by creating a folder (directory) on your blank disk named **Labs**. Copy the file **demux4chan.vhd** from the CD-ROM to the folder **Labs** on your disk.
   As you have seen in Lab 2, to create a VHDL design you must create a text file to enter the VHDL code. The text file is converted to a symbol that can be displayed in the Graphic Editor. The Graphic Editor is used to connect input and output symbols. Effectively, your VHDL code creates a new symbol in the Altera symbol library. The VHDL text file representing the DEMUX has been copied from the CD-ROM. You need only create a symbol for the VHDL DEMUX.
2. Start the **MAX+PLUS II** program. Open the file **demux4chan.vhd**. Use the section "Lab Work Procedure: VHDL Vending Machine" in Lab 2 as a guide. Review steps 1 through 5. These steps are not needed because the VHDL file was copied from the CD-ROM. You need to open the VHDL file and complete step 6. Step 6 creates the symbol. Ignore any warning message.

The text file that represents your VHDL DEMUX has been converted into a symbol that can be displayed in the Graphic Editor.

3. You will draw a diagram for the LED panel demultiplexer system. Use the Lab Work Procedure in Lab 1 as a guide, and complete steps 1 through 12. Use the project name **Lab8**. The Graphic Editor file for the LED panel demultiplexer should resemble Figure 8-14.

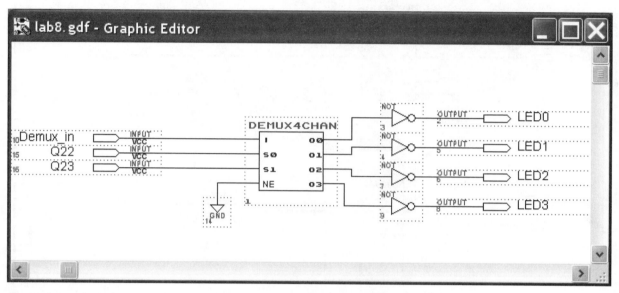

Figure 8-14 Altera "Graphic Editor": LED Panel Demultiplexer System

4. The UP board can now be programmed. Lab 1 describes all the steps that are required to program a UP board with the LED panel demultiplexer system. Use the Lab Work Procedure in Lab 1 as a guide, and complete steps 13 through 17. Ignore any compiler warning and information messages.

MAX and FLEX IC Considerations

MAX IC Design:
- Connect output **LED0**, **LED1**, **LED2**, and **LED3** to a UP board **MAX** LEDs.
- *Do not* connect inputs "Demux_in," "Q22," and "Q23" at this time. These outputs will be used to link this system to the door interface multiplexer UP board.

FLEX IC Design:
- Assign output **Door0**, **Door1**, **Door2**, and **Door3** to a UP board **FLEX** LEDs. Don't forget to turn off unused LEDs.
- Assign inputs **Demux_in**, **Q22**, and **Q23** to the **FLEX expansion header**. It will allow you to link this system to the door interface multiplexer UP board. Appendix C describes the "FLEX expansion header" in more detail. Do not connect wires to these pins at this time.

You are ready to test the operation of the LED panel demultiplexer system. Proceed to Lab Exercise 1.

LAB EXERCISES AND QUESTIONS

This section contains lab exercises that can be performed on the UP board and questions that can be answered at home. Ask your instructor which exercises to perform and which questions to answer.

Lab Exercises

Exercise 1: Test the MUX/DEMUX security system.

You will link the door interface MUX system to the LED panel DEMUX system and test the operation of the security system.

Follow this procedure to connect the two UP boards:

1. Connect a wire from **Mux_out** of the door MUX to **DEMUX_in** of the LED DEMUX.
2. Connect a wire from **Q22** of the door MUX to **Q22** of the LED DEMUX.
3. Connect a wire from **Q23** of the door MUX to **Q23** of the LED DEMUX.
4. Connect a wire to ground the **MUX UP** board to the **DEMUX UP** board. Connect a long wire from **JTAG_OUT Pin 2** of the transmitter to **JTAG_OUT Pin 2** of the receiver. Pin 2 is located at the bottom right-hand side of the JTAG_OUT connector. This ground wire is required whenever logic levels are shared between two different systems with separate power packs. Refer to Figure 8-15.

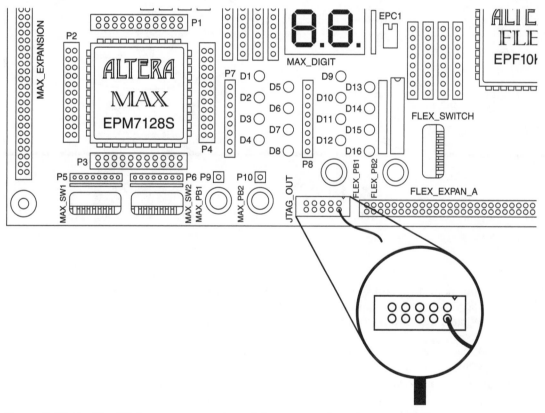

Figure 8-15 Use The JTAG connecter to ground the MUX UP board and the DEMUX UP board.

Follow this test procedure and answer questions:

a. What will a security guard see if there was an intrusion on door 1? Open door 1 and close doors 0, 2, and 3.
b. What will a security guard see if there was a simultaneous intrusion on door 1 and door 2? Open both doors 1 and 2 and close doors 0 and 3.
c. Calculate the scan rate of the security system. You may want to review the frequency division information in Lab 5.

Exercise 2: Change the security system to use Q13 and Q14.

Use the UP board to change the system, observe its operation, and answer a question. Change **Q22** and **Q23** on the **Door Interface Multiplexer** system to **Q13** and **Q14**. Recompile and reprogram (or reconfigure) the "Door Interface Multiplexer" system. Refer to Figure 8-16.

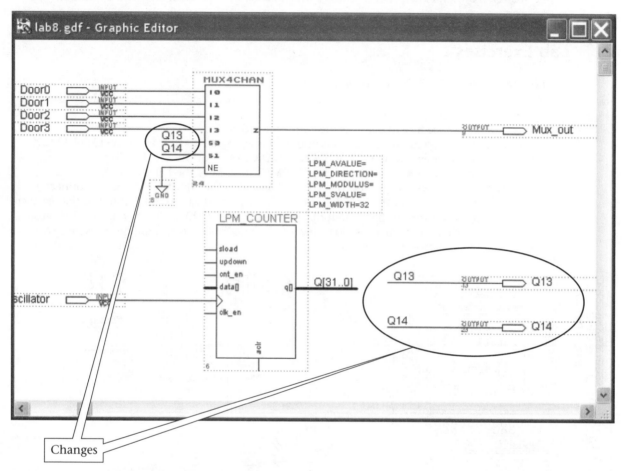

Figure 8-16 Altera "Graphic Editor": MUX/DEMUX Security System with Q13 and Q14 Replacing Q22 and Q23

    a. How does the security system work with "Q13" and "Q14?"
    b. Calculate the new scan rate.
    c. What is the advantage of a security system that uses "Q13" and "Q14" to scan the doors?
    d. What is the advantage of a security system that uses "Q22" and "Q23" to scan the doors?

Exercise 3: Change the security system to include the FAULT shown in Figure 8-20.

Use the UP board to change the system, observe its operation, and answer question 1. To add the fault shown in Figure 8-20, all you need to do is remove the "S0" wire from the Door Interface Multiplexer system and connect it to ground. Use the system to answer lab question 1.

Exercise 4: Expand the security system to monitor eight entry points.

You will need to change the code for the VHDL MUX and the VHDL DEMUX to test the operation of the new system. How many wires are required to link the MUX UP board to the DEMUX UP board?

Exercise 5: Design an advanced camera scanner system.(This exercise can be assigned as a project.)

    Review the basic camera scanner system described in Lab 7. You will design and test the operation of a camera scanner system with advanced features. The features are:

    • The system can scan 10 cameras in the **automatic mode**.

- The system uses a 10-key SPST keypad to select one camera in the **manual mode**.
- The system uses a switch to select the two modes of operation (**manual** or **automatic**). When the switch is in the manual mode, the camera is selected by the keypad. When the switch is in the automatic mode, each camera is selected sequentially for about 1 second. The manual mode allows a security guard to stop the scanning and view one camera. Refer to Figure 8-17.

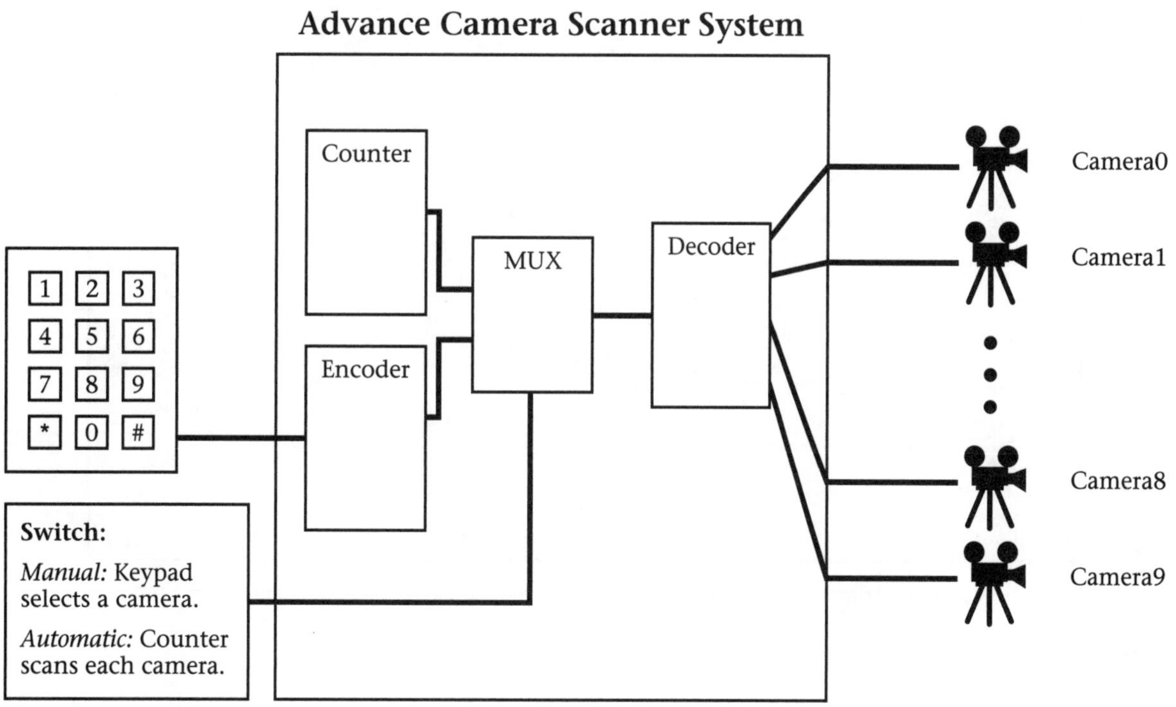

Advance Camera Scanner System

Figure 8-17 Advanced Camera Scanner System

This application requires you to use or create a variety of digital devices. Here is a partial list of devices.

- The MUX is a quad 1 of 2 MUX. It is described in the Introductory Information section. You can make it up by modifying the VHDL code for the 1 of 4 MUX or looking up the VHDL code for a quad 1 of 2 MUX on the Internet or using the symbol **74157** in the "mf" library.
- The counter is a BCD counter. You can make up a BCD counter by using the VHDL BCD counter from Lab 5 or using the symbol **lpm_counter** in the "mega_lpm" library. The **lpm_counter** symbol has options to create a BCD counter. You must set **LPM_MODULUS** to **10** and **LPM_WIDTH** to **4**.
- The keypad encoder can be made up by using the VHDL encoder from Lab 7 or using the symbol **74147** in the "mf" library.
- The decoder can be made up by modifying the VHDL decoder from Lab 7 or using the symbol **7445** in the "mf" library.

NOTE: *The symbols "74157," "74147," and "7445" are a carryover from the old TTL IC technology. You can figure out how they operate by searching the Web for TTL data sheets, or you can analyze their internal construction by double-clicking on the symbol when they are placed on the worksheet in the "Graphic Editor" window. It should also be noted that you could make up the devices by using the "lpm" symbols found in the "mega_lpm" library. Ask the instructor for guidance.*

Exercise 6: Use the Altera simulator to test the security system's door interface multiplexer.
(This exercise can be done at home.)

Use the MAX+PLUS II "Graphic Editor" to make changes to the **lab8.gdf** file for the door interface multiplexer system. The changes will simplify the system and allow you to use the simulator to test its operation.

1. Add two output symbols. Name one **Q0** and the other **Q1**.
2. Change the name of input **S0** and **S1** from **Q22** and **Q23** to **Q0** and **Q1**.

The changes are shown in Figure 8-18.

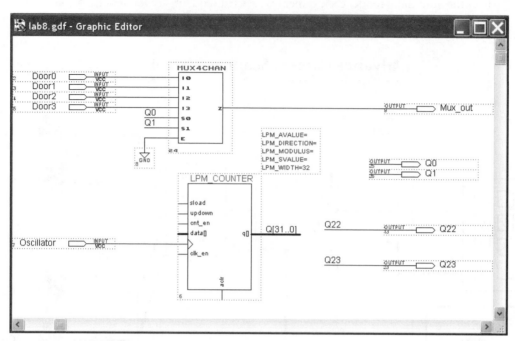

Figure 8-18 Altera "Graphic Editor": Simulator Changes for the Security System Door Interface Multiplexer System

Use Appendix F as a guide to run a simulation of the random number generator system you have just created. The changes may give warnings when you use "Save, Compile & Simulate." These warnings can be ignored. The "Waveform Editor" window for the simulation should resemble Figure 8-19.

Figure 8-19 Waveform Editor Window for the Security System Door Interface Multiplexer System

Set the **End time** to 1.0us. The "Oscillator" uses an "Overwrite clock period" = 40.0ns. If the "Clock Period" entry box is grayed out (not selectable), then you should exit the "Overwrite Clock" window, select the **Options** menu, and remove the check mark beside the menu item "Snap to Grid." Return to the "Overwrite Clock" window and you will be able to set the clock period.

a. Open only "Door0" and run the simulation. To open "Door0" you can right click on it and then select "Overwrite" then "High(1)." Run the simulation and explain the response you observe at the "Mux_out" output.

b. Group the "Q" outputs into a single node. Close "Door0" and open both "Door1" and "Door2." To close "Door0" you can right click on it and then select "Overwrite" then "Low(0)." To open "Door1" you can right click on it and then select "Overwrite" then "High(1)." Repeat this process to open "Door2." Run the simulation and explain the response you observe at the "Mux_out" output.

c. Open all doors and run the simulation. Explain the response you observe at the "Mux_out" output.

Lab Questions

1. A 4-door security system breaks down. The fault is shown in Figure 8-20. Fill in the table to show the system response to the fault. Explain how the fault would confuse a security guard using the system. Does the system partially work?

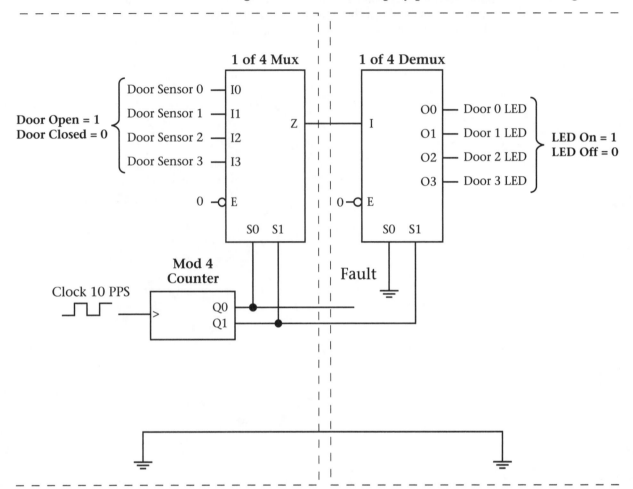

Figure 8-20 Faulty Security System for Question 1

Counter		Mux		Demux		Door #	LED #
Q1	Q0	S1	S0	S1	S0	Selected	Selected
0	0						
0	1						
1	0						
1	1						

2. Lab 2 describes a drill machine system. The description can be found in the Lab Exercise section near the end of the lab. Use a single multiplexer to design a logic gate system for the drill machine system. Using a multiplexer as a logic gate system is described in the Introductory Information section of this lab.

3. Show how to combine three 1 of 4 multiplexers to create a 1 of 8 multiplexer. Each 1 of 4 multiplexer has an active low enable input.
4. Study the quad 1 of 2 MUX system shown in Figure 8-21.

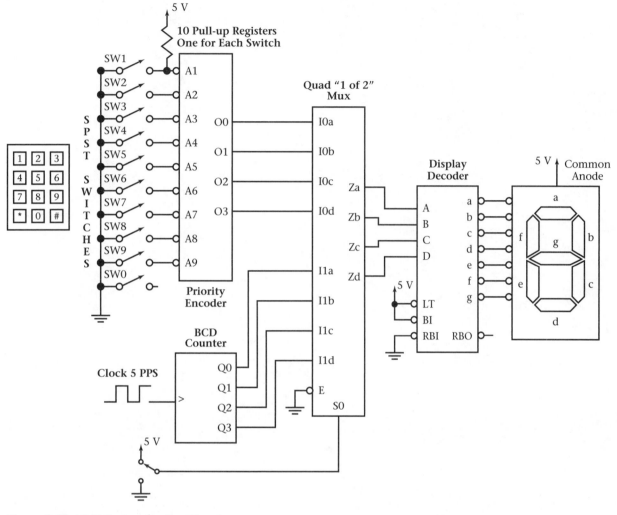

Figure 8-21 MUX System for Question 4

    a. Explain what would appear on the digital display if the switch is in the 0 position.
    b. Explain what would appear on the digital display if the switch is in the 1 position.
5. Write the code for a VHDL "1 of 8" MUX with an active high enable.
6. Write the code for a VHDL Quad "1 of 2" MUX with an active low enable.

CONCLUSIONS

- Multiplexers are data selectors.
- Demultiplexers are data distributors.
- A MUX/DEMUX security system reduces the amount of wiring from the door sensors to the LED panel.
- The MUX/DEMUX security system demonstrates time division multiplexing (TDM).
- A single multiplexer can be used to construct an entire logic gate system.

Lab 9

Matrix Keypad Encoder System

Equipment

Altera UP board
Book CD-ROM
Spool of 24 AWG wire
Wire cutters
Floppy Disk with Lab 3 debounce files
4 x 4 Matrix keypad (optional)

Objectives

Upon completion of this lab, you should be able to:

• Build a matrix keypad encoder system.

INTRODUCTORY INFORMATION

There are two types of keypads. One uses individual SPST switches and the other uses switches that are arranged in a matrix of rows and columns. Lab 7 described how to encode an SPST keypad. This lab will show you how to build an encoding system for a matrix keypad.

Matrix Keypad Theory

A matrix keypad has pushbutton switches arranged in rows and columns. Refer to Figure 9-1.

A matrix keypad uses pushbuttons that are similar to buttons you find on a telephone. You press and hold them down to connect a row to a column. You release them and they spring back up to disconnect the row and the column. For example, Figure 9-2 shows what happens when you press and hold down the "key 0."

Each button pressed connects a unique row number to a unique column number. When the buttons on a matrix keypad are released, the rows and columns are disconnected from each other. The advantage of a matrix keypad is the reduction in connection points. The 16-key

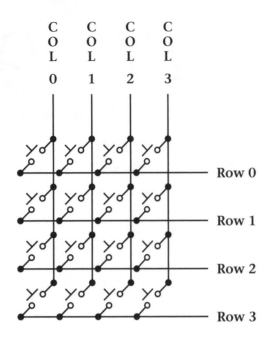

Figure 9-1 Matrix Keypad

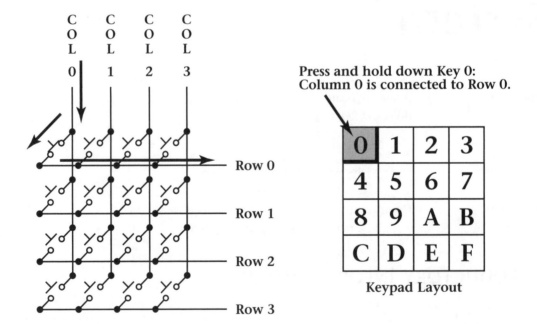

Figure 9-2 Matrix Keypad: Press And Hold Key 0.

keypad shown in Figure 9-1 requires eight connections (four rows and four columns). A computer keyboard, which has about 100 keys, can be connected in matrix format that requires as little as 20 connection points (10 rows × 10 columns). The disadvantage of using a matrix keypad is that the encoder system is more complex to build.

Matrix Keypad Encoder: Operation Summary

Figure 9-3 shows the diagram for a matrix keypad encoder system.

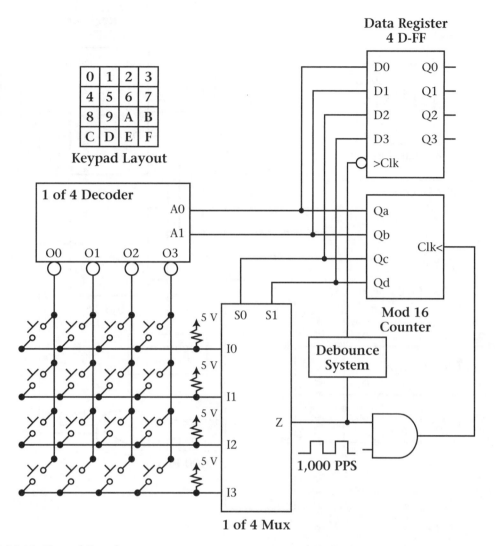

Figure 9-3 Matrix Keypad Encoder

Here is quick summary of the operation of the matrix keypad encoder. A detailed description is presented later. The matrix keypad encoder is made up of several devices that work together to store the number of the key pressed into the "data register (4-D FF)." The heart of the system is the **mod 16 counter**. The **counter** generates **16** key numbers. The **"1 of 4"** **decoder** and the **"1 of 4"** MUX use these numbers to scan the matrix keypad to identify which key is pressed. Once the key is identified, it is stored in the **data register**. The number remains stored in the "data register" until the next key is pressed.

Matrix Keypad Encoder: Operational Details

Case 1: No Keys Pressed:

Figure 9-4 shows the operation of the matrix keypad encoder system when none of the keys are pressed.
The MUX:
When no keys are being pressed, the "rows" of the keypad will not be connected to any of the "columns." The four "pull-up resistors" will make the four "input channels" of the "1 of 4 MUX" = "1". Regardless of which channel is selected, "Z" of the "MUX" is "1."

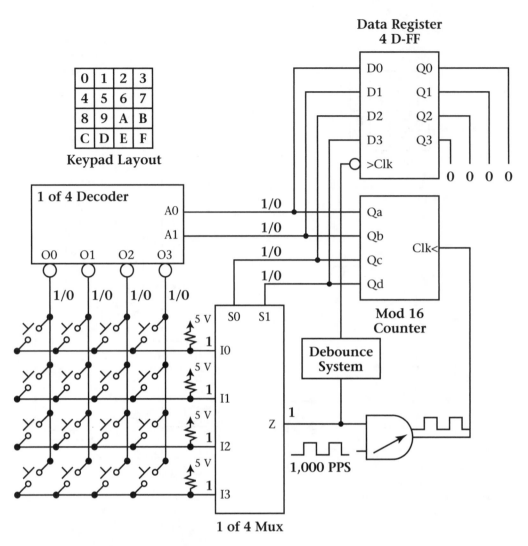

Figure 9-4 Matrix Keypad Encoder: No Keys Pressed

The counter:

With "Z = 1," the "AND" gate passes the "1,000 PPS" pulse to the clock of the "counter." The "counter" cycles quickly and repeatedly from "0 to 15." The label "1/0" at "Qa," "Qb," "Qc," and "Qd" represents the continuous cycling at the output of the "counter."

The decoder:

The two least significant bits (LSB) of the "counter" generate a continuous cycle of "0, 1, 2, and 3" at the input of the "decoder." The output of the "decoder" will have one output at "0" for each count. The result is a "0" is moved from "O0" toward "O3" for each cycle of the "counter." The "0" grounds each column of the keypad successively. The "0" is moved from "column 0" toward "column 3" and then the entire process is repeated. The label "1/0" at "O0", "O1," "O2," and "O3" represents the continuous movement of the "0" at the output of the "decoder." The term **scanning the keypad columns** is used to describe this action. The "counter" uses the "decoder" to scan the columns of the "keypad."

The MUX revisited:

The two most significant bits (MSB) of the counter generate a continuous cycle of "0, 1, 2, and 3" at the "select inputs" of the "MUX." Each "MUX" input channel is passed to "Z" while the columns of the keypad are being scanned. The term **scanning the keypad rows** is used to describe this action. With no keys pressed, all input channels are "1"; thus, "Z" is always "1" regardless of which row is selected.

The data register:

The initial condition of the "data register" is "Q3 = Q2 = Q1 = Q0 = 0." With "Z = 1," the negative edge triggered clock input of the "data register" is not asserted. The "data register" is inactive and continues to store "0000" (#0).

Conclusion:

The "counter" counts continuously, the "decoder" continuously moves a "0" across the columns of the "keypad," the "MUX" scans the rows and makes "Z = 1," and the "data register" is not used because the clock is not asserted.

Case 2: Press and Hold Down Key "5":

Figure 9-5 shows the operation of the matrix keypad encoder system when "key 5" is pressed.

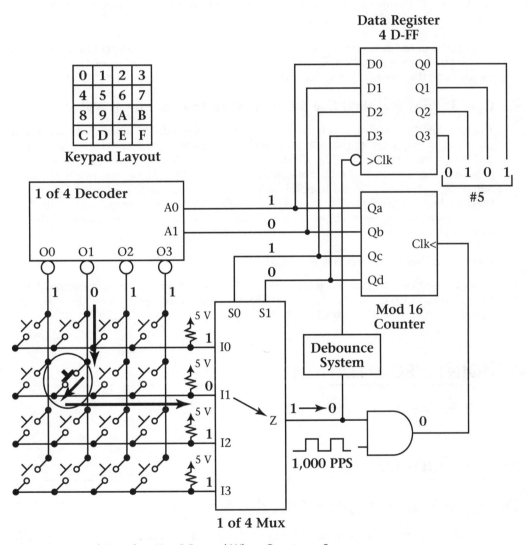

Figure 9-5 Matrix Keypad Encoder: Key 5 Pressed When Counter = 5

Assume that "key 5" is pressed when the "counter = 5." It is important to understand that unless "Z" changes from "1 to 0," the counter keeps cycling and the system responds as if there are no keys pressed. The "0" that is transferred to "Z" comes from the "decoder." The key pressed connects a column and a row, which connects the "decoder" output to the "MUX" input. The analysis assumes "key 5" is pressed when the "counter =5." Later the system will be reanalyzed with the "counter" at a number other than "5."

The decoder:

With the "counter = 5," the two LSB of the "counter" place the number "1" at the input of the "decoder." The "decoder" will output a "0" (ground) at "O1." The "0" is sent to the "MUX" input channel "I1" through the pressed "key."

The MUX:

Input channel "I1 = 0" because the "keypad" connects the "MUX" row to the "decoder" column. The two MSB of the "counter" place the number "1" at the select inputs of the "MUX." The "MUX" selects channel "I1" and makes "Z = 0."

The counter:

When "Z" changes from "1 to 0," the "AND" gate blocks the "1,000 PPS" pulse from the clock input of the "counter." The "counter" stops and freezes on the number "5."

The data register:

When "Z" changes from "1 to 0," it creates a negative edge transition that clocks the "data register" and stores the number "5."

Conclusion:

The "counter" stops on "5," the decoder grounds the input channel of the "MUX" through the "keypad" column and row connection, the "MUX" changes "Z" from "1 to 0," which blocks the clock to the "counter" and clocks the "data register." Key number "5" is stored.

Case 3: Press and Hold Key 5 Down When the Counter is *Not* at 5:

The analysis of the system is simple. If the "counter" is not at "5" when "key 5" is pressed, then "Z = 1." The only number that will ground the column to a row and makes "Z = 0" is the number "5." A number other than "5" may ground a column that is not connected to the key being pressed or select a row that does not contain the grounded column. Regardless, the result is the same; "Z = 1" and the system responds as though no key is pressed. With "Z = 1" the counter keeps cycling at "1,000 PPS." Fortunately, at "1,000 PPS," it does not take very long for the "counter" to cycle to the number "5" and change "Z" to "0." The term **scanning the keypad** is used to describe the counting action while "Z = 1."

Case 4: Other Keys on the Keypad:

Other keys on the "keypad" each have their own unique "counter" number that will cause "Z" to change from "1 to 0." The "counter" scans the "keypad" until the correct "decoder" and "MUX" combination makes "Z" change from "1 to 0."

POWERPOINT PRESENTATION

Use PowerPoint to view the Lab 09 slide presentation.

LAB WORK PROCEDURE

Design a Matrix Keypad Encoder System

A matrix keypad will be connected to an encoder system. You will need to use the information in the introductory section of this lab to design the system. Unlike most labs in this book, you will not be given a MAX+PLUS II diagram to follow as a guide. This makes the design a little more challenging. However, you will be given a suggested list of Altera symbols to use. If you don't have a 4×4 matrix keypad, you can complete the lab by connecting the four row inputs and four column outputs as follows:

- Connect four row inputs to four MAX switches. Flip the switches up to 1. This will simulate the pull-up resistors at the inputs of the MUX.
- Connect one end of four wires to the four columns. Leave the other ends open.
- To simulate pressing a key, you can touch one of the four open column wires to a row wire. Touch the row wire where it terminates at the MAX switch.

Using a matrix keypad is preferable.

1. Draw a diagram for the "matrix keypad encoder system." Use the Lab Work Procedure in Lab 1 as a guide and complete steps 1 through 12. Use the disk that contains the Lab

3 debounce files, in folder **Labs** to create a project named **Lab9**. Here is a list of suggested Altera symbols in the "mega_lpm" library you may want to use:
- Mod 16 counter: **LPM_COUNTER** with **LPM_MODULUS = 16** and **LPM_WIDTH = 4**.
- Multiplexer: **LPM_MUX** with **LPM_SIZE = 4** and **LPM_WIDTH = 1**.
- Decoder: **LPM_DECODE** with **clock = aclr = unused** and **LPM_WIDTH = 2**.
- Data register: **LPM_DFF** with **shiftin = shiften = shiftout = enable = aclr = unused** and **LPM_WIDTH = 4**.

2 The UP board can now be programmed. Lab 1 describes all the steps that are required to program a UP board with the keypad encoder system. Use the Lab Work Procedure in Lab 1 as a guide, and complete steps 13 through 17. Ignore any compiler warning and information messages.
- Connect the four rows (with pull-up resistors) and four columns of the matrix keypad to the UP board (or if you don't have a keypad, refer to the description given at the start of the lab work procedure).

MAX and FLEX IC Considerations:

MAX IC design:
- Assign input **Oscillator** to pin number **83**. A PCB trace connects the 25,175,000 PPS oscillator to pin 83 of the MAX IC. Refer to Lab 1, step 13_{FLEX} for assistance in assigning an input symbol to a pin number.

FLEX IC design:
- Assign input **Oscillator** to pin number **91**. A PCB trace connects the 25,175,000 PPS oscillator to pin 91 of the FLEX IC.
- Assign row and column input output symbols to the FLEX expansion header. Appendix C describes the FLEX expansion header in more detail.

Lab Exercises and Questions

This section contains lab exercises that can be performed on the UP board and questions that can be answered at home. Ask your instructor which exercises to perform and which questions to answer.

Lab Exercises:

Exercise 1: Test the matrix keypad encoder system.
- a. Make sure the system is fully operational before proceeding. Test each key individually.
- b. Use the UP board to change the system, observe its operation, and answer a question. How well does the system work if the debounce system were bypassed? To bypass the debounce system, connect **Z** directly to the data register clock. Try pressing keys. You may need to quickly and repeatedly tap a key to see any effect of switch bounce. Be sure to reconnect the debounce system after this test is complete.

Exercise 2: Add a digital display to the matrix keypad system.
Use the UP board to change the system and observe its operation. Add the **display decoder** from Lab 6 to the **matrix keypad** system. The system will display the key number on a digital display.

Exercise 3: Test the operation of the system using an active high "1 of 4" column decoder.
Use the UP board to change the system, observe its operation, and answer a question. Change the current active low "1 of 4" column decoder and make it an active high "1 of 4" column decoder. How does the system work? Explain why it works this way.

Exercise 4: Design a two-key storage system.
Use the UP board to change the system and observe its operation. Add a second **display decoder** and a second **data register** to create a system capable of storing and displaying two separate key numbers. The first key is displayed on the first **digital display** and the second key is displayed on the second digital display. You will need to use some creative thinking to control the clock inputs of the two data registers. You do not want one key to be displayed twice. Be sure to use the **ripple**

blanking feature to turn off leading zeros. For example, if the numbers 07 are stored, they will be displayed as the number 7 without a leading 0.

Lab Questions

1. Explain how the matrix keypad encoder system would respond if KEY 0 and KEY 2 were both pressed and held down at exactly the same time. Which key number will likely get stored? Explain.
2. Explain how the matrix keypad encoder system would operate with the following error: Disconnect "A0" of the decoder from "Q0" of the counter. Connect "A0" of the decoder directly to "Vcc" (logic 1). Be sure to list examples of keys pressed that would work and some that would fail.
3. Connect the matrix keypad to the "1 of 4" decoder and the "1 of 4" mux in order to change the keypad layout. Refer to Figure 9-6.

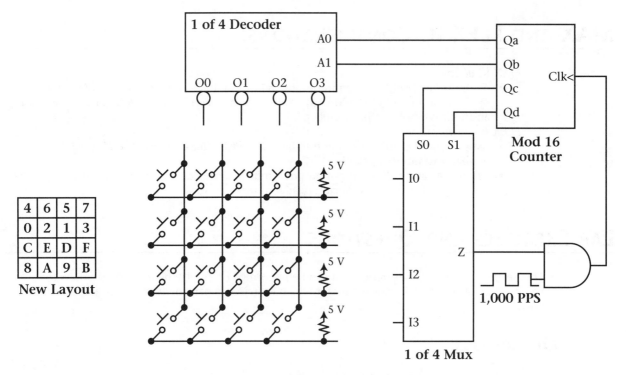

Figure 9-6 Partial Matrix Keypad Encoder Diagram for Question 3

Conclusions:

*   A matrix keypad has pushbutton switches arranged in rows and columns. You press and hold them down to connect a row to a column.
*   A matrix keypad reduces the number of connection points. A 16-key keypad has eight connection points.
*   A matrix keypad requires a more complicated encoder system than an SPST keypad.

Lab 10

Arithmetic Systems

Equipment
    Altera UP board

    Book CD-ROM

    Spool of 24 AWG wire

    Wire cutters

    Blank floppy disk

Objectives
    Upon completion of this lab, you should be able to:

- Design a system that can be used to add and subtract binary numbers.
- Convert signed numbers using 2s complement notation.

INTRODUCTORY INFORMATION

Arithmetic systems can be used to add, subtract, multiply, and divide binary numbers. A system that can add and subtract binary numbers is studied in this lab.

Binary Addition

To understand the adder/subtractor system you must begin by understanding binary hand addition. Figure 10-1 shows how to add 3 + 1 using 4-bit hand addition.

An adder system must replicate the hand addition process. It must be able to add 2 bits to a **carry** and generate a sum and a **carry out**. A **full adder** is the name given to a digital element that can do this. Figure 10-2 shows a full adder. A 4-bit adder system is made up of four full adder elements. Figure 10-3 shows how 3 + 1 is added using the 4-bit adder system.

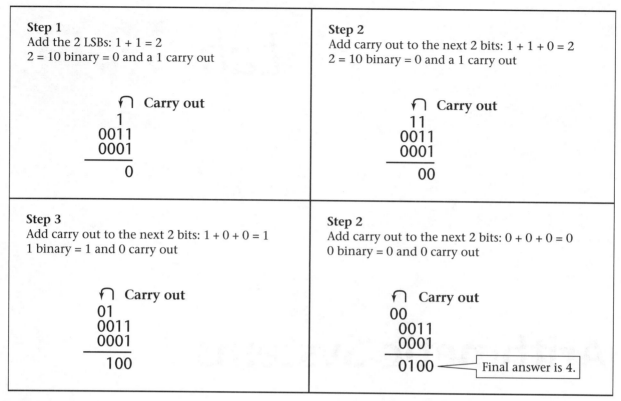

Step 1
Add the 2 LSBs: 1 + 1 = 2
2 = 10 binary = 0 and a 1 carry out

⌐ **Carry out**
1
0011
0001
———
0

Step 2
Add carry out to the next 2 bits: 1 + 1 + 0 = 2
2 = 10 binary = 0 and a 1 carry out

⌐ **Carry out**
11
0011
0001
———
00

Step 3
Add carry out to the next 2 bits: 1 + 0 + 0 = 1
1 binary = 1 and 0 carry out

⌐ **Carry out**
01
0011
0001
———
100

Step 2
Add carry out to the next 2 bits: 0 + 0 + 0 = 0
0 binary = 0 and 0 carry out

⌐ **Carry out**
00
0011
0001
———
0100 ◁— Final answer is 4.

Figure 10-1 4-Bit Binary Hand Addition

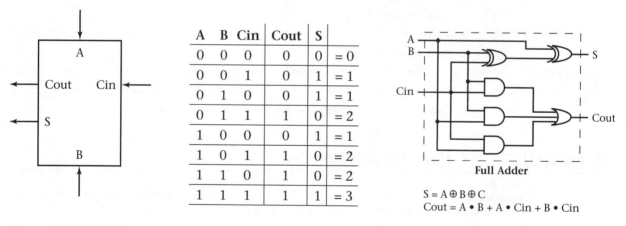

A	B	Cin	Cout	S	
0	0	0	0	0	= 0
0	0	1	0	1	= 1
0	1	0	0	1	= 1
0	1	1	1	0	= 2
1	0	0	0	1	= 1
1	0	1	1	0	= 2
1	1	0	1	0	= 2
1	1	1	1	1	= 3

Full Adder

$S = A \oplus B \oplus C$
$Cout = A \cdot B + A \cdot Cin + B \cdot Cin$

Figure 10-2 Full Adder

The Altera symbol library includes an 8-bit adder symbol called **8fadd**. It is equivalent to eight full adder elements. Refer to Figure 10-4.

Cin connects to full adder **A1/B1**. **Cout** connects to full adder **A8/B8**. All of the other Cin and Cout links are internally connected.

Fundamental Concepts for Subtraction: 2s Complement Notation

A subtraction system operates on the principle that subtraction is equivalent to the addition of negative numbers. In other words, A – B is equivalent to A + (–B), where *A* represents one number, the "+" symbol represents the adder system, and *(–B)* represents a negative number that

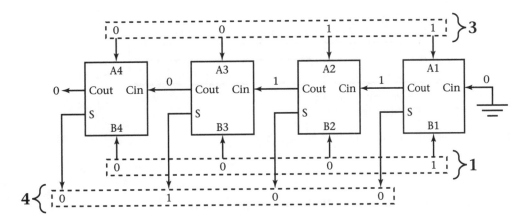

Figure 10-3 4-Bit Adder

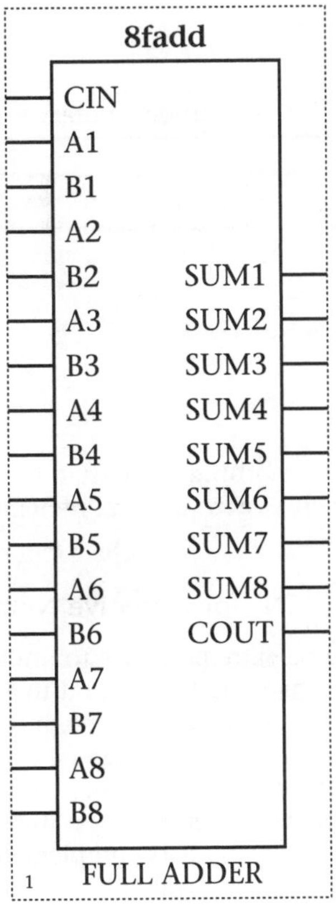

Figure 10-4 Altera 8fadd Symbol

is sent to the adder system along with *A*. This principle allows one digital system, the adder, to do both operations. To accomplish this, digital systems must use a notation to represent **negative numbers**. The signed number notation is called **2s complement notation (2CN).**

2s Complement Notation (2CN)

2CN is a signed number notation that is used to represent positive and negative numbers. The most significant bit (MSB) of a 2CN number is used to indicate the sign of the number. If the MSB is "0," then the number is positive. If the MSB is "1", then the number is negative. Figure 10-5 shows how you would use 2CN to represent an 8-bit number.

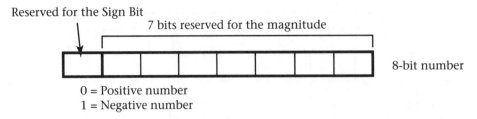

Reserved for the Sign Bit

7 bits reserved for the magnitude

8-bit number

0 = Positive number
1 = Negative number

Figure 10-5 2s Complement Notation for an 8-Bit Number

2s Complement Notation (2CN) for Positive Numbers

To represent positive numbers using 2CN, you must:
1. Make the MSB = 0 to indicate the number is positive.
2. Use the remaining magnitude bits to represent the number in binary.

Figure 10-6 shows how you would use 2CN to represent the number 8.

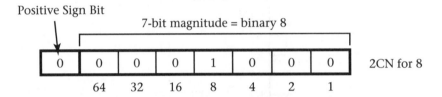

Positive Sign Bit

7-bit magnitude = binary 8

| 0 | 0 | 0 | 0 | 1 | 0 | 0 | 0 | 2CN for 8
| | 64 | 32 | 16 | 8 | 4 | 2 | 1 |

Figure 10-6 2s Complement Notation for the Number 8

2CN for positive numbers is nothing more than binary notation with a sign bit set to 0. This makes 2CN for positive numbers easy to use and remember. Negative numbers, on the other hand, are a little more complex.

2s Complement Notation (2CN) for Negative Numbers

Negative 2CN numbers would be extremely easy to understand if all you had do is change the sign bit to a 1. Unfortunately, changing the sign bit to a 1 would create a binary number that would not yield subtraction by using an adder system. Remember, the purpose of 2CN is to represent a number that will generate subtraction by using an adder system.

To represent negative numbers using 2CN, you must:
1. Represent the number as a *positive* 2CN number.
2. Invert all the bits. This creates a 1s complement notation (1CN) number.
3. Add 1 to the 1CN number to create a 2CN number.

Figure 10-7 shows how you would use 2CN to represent the number –8. The final result shows that the 2CN for –8 is a binary pattern that is not immediately recognizable. Not many people can look at 11111000 and say that this looks like the binary pattern for –8. On the other hand, most people with knowledge of binary numbers can look at 00001000 and say that this looks like the binary pattern for +8 because 2CN for positive numbers is binary notation. The binary pattern for –8, however, does allow an adder to do subtraction. Figure 10-8 shows what happens when you add +8 to –8.

2CN does allow an adder system to do subtraction as long as you remember to ignore the carry out of the MSB. In digital systems, the carry out of the MSB is not connected to anything in the system and is not used.

Valid Range for 2CN Numbers

An arithmetic system designer must choose the number of bits the system will use. This choice can critically affect the operation of the system. Choosing too few bits will result in an arithmetic system that many users will find ineffective. For example, assume a designer would like to build a

Procedure to convert the number −8 into a 2CN number.

Step 1. Write 2CN for +8.

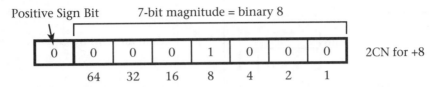

Step 2. Invert all bits. 1CN number.

1	1	1	1	0	1	1	1

Step 3. Add 1 to the 1CN number.

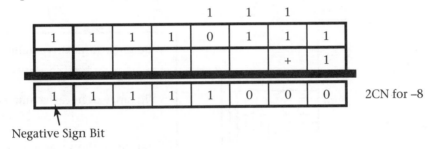

Figure 10-7 2s Complement Notation for the Number −8

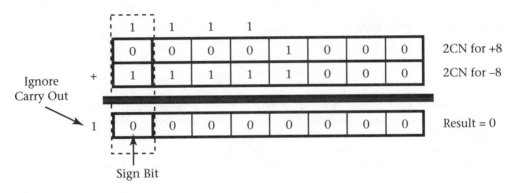

Figure 10-8 Adding +8 to −8 Using 2CN

simple calculator system to be used by elementary school children. The children will use the calculator to check the results of hand addition and hand subtraction problems.

Case 1: Design a 4-Bit Arithmetic System

A 4-bit arithmetic system uses 1 bit for the sign of the number and 3 bits for the magnitude. Figure 10-9 shows the valid range of numbers for a 4-bit system.

Case 2: Design an 8-Bit Arithmetic System

An 8-bit arithmetic system uses 1 bit for the sign of the number and 7 bits for the magnitude. Figure 10-10 shows the valid range of numbers for an 8-bit system.

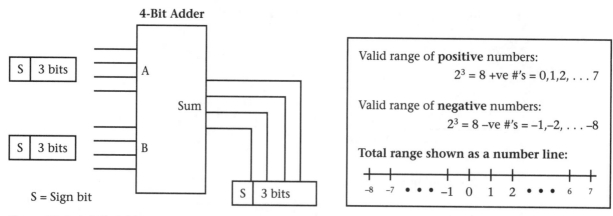

Figure 10-9 A 4-Bit Arithmetic System

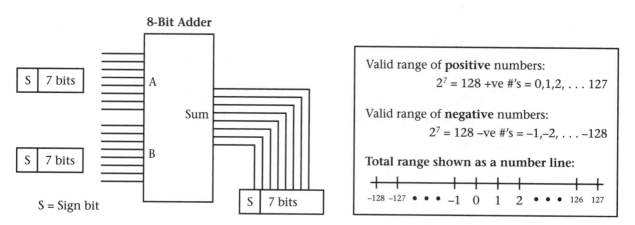

Figure 10-10 An 8-Bit Arithmetic System

The two cases clearly show the impact of a designer's choice for the number of bits the arithmetic system will use. The 4-bit system can only generate valid mathematical operations on a range of 16 numbers. This limits the range of math operations from −8 through +7. Not many school children need a calculator system to verify hand operations limited to this range. The 8-bit system fairs better because its valid range of 256 numbers covers a span of −128 to 127. For most arithmetic systems, the more bits you use, the more beneficial the system is to the user.

Addition Using 5-Bit 2CN Numbers and Arithmetic Overflow

To demonstrate how 2CN addition works, several 5-bit numbers will be added together. The valid range of 5-bit numbers is shown in Figure 10-11.

Case 1: Positive Number and a Smaller Negative Number [9 + (−4)]

Figure 10-12 shows how the number 9 is added to the number −4 to generate +5.

Case 2: Positive Number and a Larger Negative Number [−9 + 4]

Figure 10-13 shows how the number −9 is added to the number 4 to generate −5.

The final answer is "1 1011." The sign bit seems correct because it is "1" indicating a negative number. The magnitude portion of the answer is "1011". How can you verify that this is the 2CN number for "−5"? There is a technique that can be used to verify the answer. The verification technique is based on this theory: *A positive 2CN number converts to its magnitude equivalent negative 2CN number, and a negative 2CN number converts to its magnitude equivalent positive 2CN number.* You have seen that a negative 2CN number is derived from a positive 2CN

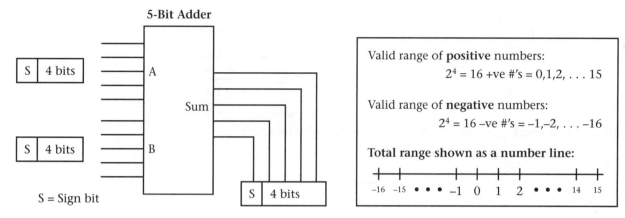

Figure 10-11 Valid Range of 5-Bit Numbers

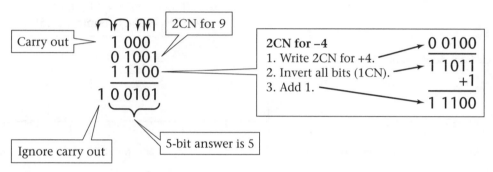

Figure 10-12 5-Bit Addition of "9 +[–4]"

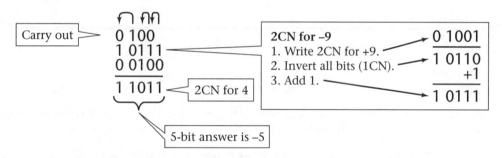

Figure 10-13 5-Bit Addition of "[–9] + 4"

number that has been inverted and then 1 has been added to it. What you are about to find out is that if you start with a negative 2CN number and then you invert it and then add 1, it will revert back to a positive 2CN number. The **invert add 1** process can be used to flip positive numbers to negative numbers and flip negative numbers to positive numbers. Figure 10-14 shows how you can check the answer.

Case 3: Two Positive Numbers That Cause an Addition Error [9 + 8]

Figure 10-15 shows what happens when the number 9 is added to the number 8. The error illustrated in Figure 10-15 is called **arithmetic overflow**. *Arithmetic overflow* is an error that can occur if two positive 5-bit numbers added together generate a sum that exceeds the valid range. Remember, the valid range of 5-bit numbers includes –16 to +15. Seventeen (9 + 8) is not part of this range. Arithmetic overflow occurs when 2 positive numbers are added and their sum generates a negative number. On a number line, it is equivalent to exceeding the end of the line

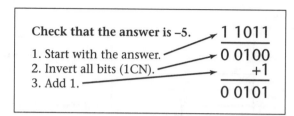

Figure 10-14 Check the answer for the example shown in Figure 10-13.

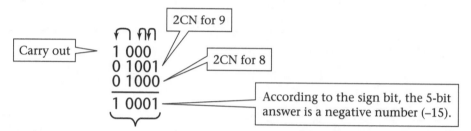

Figure 10-15 5-Bit Addition of "9 + 8"

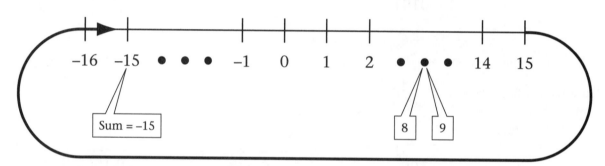

Figure 10-16 Using a number line to show arithmetic overflow "9 + 8"

and overflowing or wrapping back around to the end of the number line where the negative numbers are located. Figure 10-16 illustrates this type of arithmetic overflow.

Case 4: Two Negative Numbers That Cause an Addition Error [–9 + (–8)]

Figure 10-17 shows what happens when the number –9 is added to the number –8. Figure 10-17 illustrates a second way to generate an arithmetic overflow error. Arithmetic overflow can occur if two negative 5-bit numbers added together generate a sum that exceeds the valid range. Remember, the valid range of 5-bit numbers includes –16 to +15. Negative 17 (–9 + [–8]) is not part of this range. Arithmetic overflow occurs when two negative numbers are added and their sum generates a positive number. On a number line, it is equivalent to exceeding the end of the line and overflowing or wrapping back around to the end of the number line where the positive numbers are located. Figure 10-18 illustrates this type of arithmetic overflow.

Case 5: Numbers That *Never* Cause an Addition Error

Figure 10-19 uses a number line to show that arithmetic overflow never occurs if the numbers being added together come from each side of the number line. One number is positive and the other is negative.

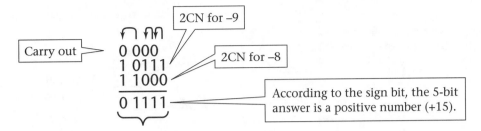

Figure 10-17 5-Bit Addition of "–9 + (–8)".

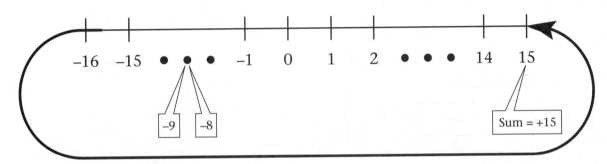

Figure 10-18 Using a number line to show arithmetic overflow "[–9] + (–8)"

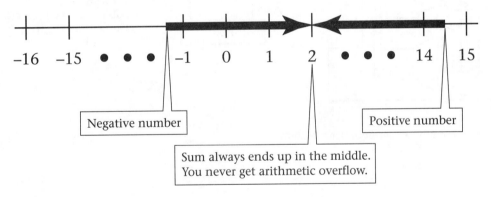

Figure 10-19 Using a number line to show when arithmetic overflow does not occur

Arithmetic Overflow Detection System

An arithmetic overflow detection system need only monitor the status of the sign bits of each number and the sign bit of the sum. Figure 10-20 shows a logic gate system that can be used to detect overflow.

Combination Adder / Subtractor System

Figure 10-21 shows a system that can be used to add or subtract 5-bit numbers. The XOR gates will pass the "B" number to the adder system with "Cin = 0" when in the add mode. The XOR gates will pass the inverted "B" number to the adder system with "Cin = 1" when in the subtract mode. The XORs perform a 1CN conversion and "Cin = 1" adds 1 to the 1CN conversion to complete the 2CN conversion.

Figure 10-22 demonstrates how this system is used to add 4 + (–2).

Figure 10-23 demonstrates how this system is used to subtract 4– (–2).

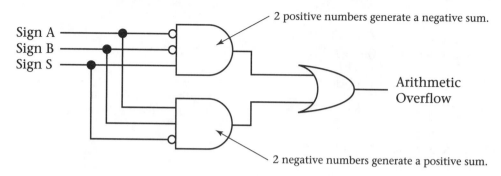

Figure 10-20 Arithmetic Overflow Detection System

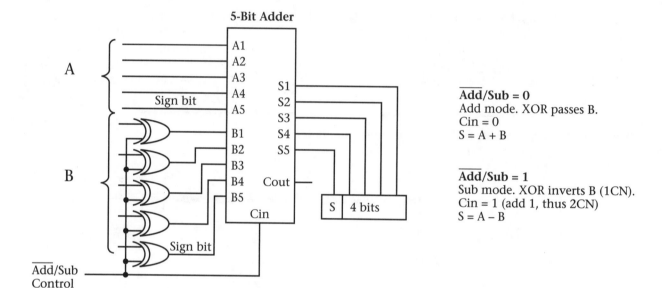

Figure 10-21 Combination Adder/Subtractor System

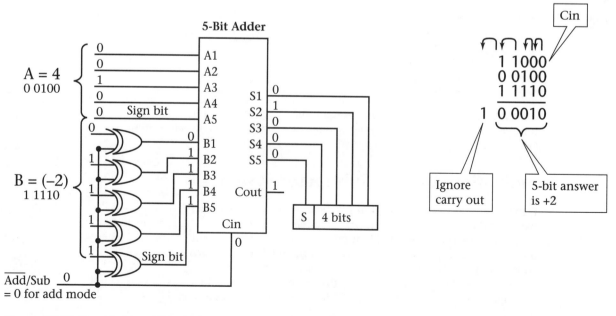

Figure 10-22 Combination Adder/Subtractor System: Add 4 + (−2)

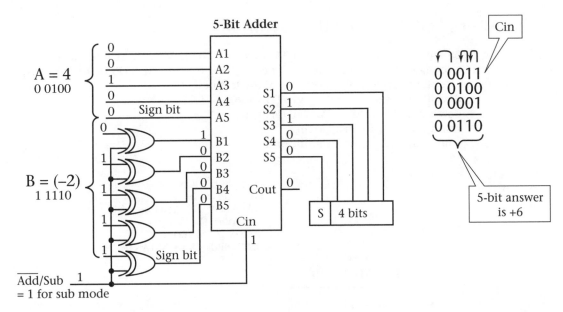

Figure 10-23 Combination Adder/Subtractor System: Subtract 4 – (–2)

2CN Special Case: The Largest Negative Number in the Valid Range

The largest negative number in a range of valid 2CN numbers has a special characteristic. It is always written with a sign bit of 1 and the magnitude bits are all zeros. This characteristic is independent of the number of bits used to represent the number. For example, the largest negative 5-bit number is –16. It is written as 1 0000. The largest negative 8-bit number is –128. It is written as 1 0000000.

You have learned that if you start with a negative 2CN number and then you invert it and then add 1, it will generate a positive 2CN number. Figure 10-24 shows what happens when you try to verify the value of –16.

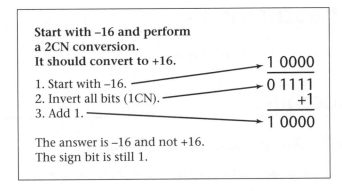

Figure 10-24 2CN check of largest negative number

What happened? The 2CN conversion generated the same binary pattern for –16. One way to look at it is to realize that there is no such thing as a 5-bit number for +16 and the 2CN conversion should not work. Remember, the largest positive 5-bit number is 15. Negative 16 is the only negative number that does not have a magnitude equivalent positive number. This characteristic is true for all the largest negative numbers no matter how many bits are used.

The Altera "LPM_ADD_SUB" Symbol

Figure 10-25 shows the Altera **LPM_ADD_SUB** symbol. The LPM_ADD_SUB has many features and can be configured many ways. You will need to read over the **Max+Plus II** online help manual to see all the details. Here are some of the features:

- **dataa[]**: input port A, **LPM_WIDTH** wide
- **datab[]**: input port B, **LPM_WIDTH** wide
- **result[]**: dataa[] + datab[] + cin
- **cout:** Carry out of most significant bit

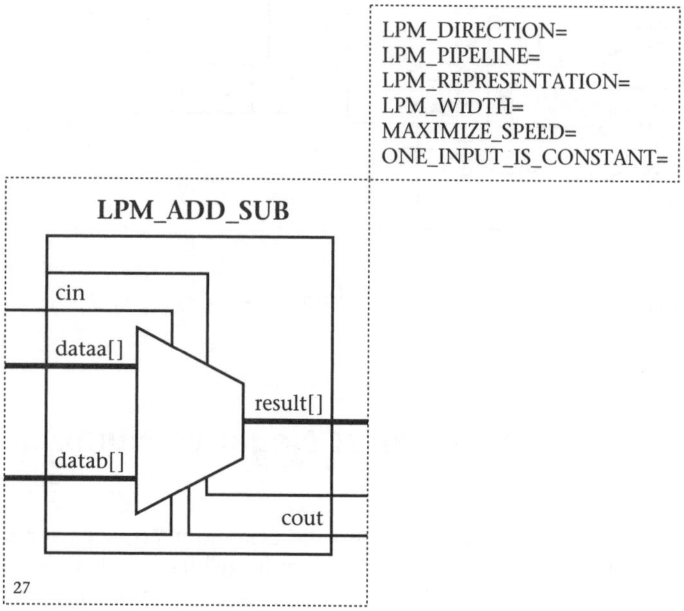

Figure 10-25 Altera "LPM_ADD_SUB" Symbol

POWERPOINT PRESENTATION

Use PowerPoint to view the Lab 10 slide presentation.

LAB WORK PROCEDURE

Design a 4-Bit Adder/Subtractor System

You will build a combination adder/subtractor system. It will generate results for 4 bit signed numbers.

1. You will draw a diagram for the 4-bit adder/subtractor system. Use the Lab Work Procedure in Lab 1 as a guide, and complete steps 1 through 12. Begin by creating a folder (directory) on your blank disk named **Labs**. Use the project name **Lab10**. The adder symbol in the "mega_lpm" library is called **LPM_ADD_SUB**. The XOR gate symbol in the "prim" library is called **XOR**. The NOT gate symbol in the "prim" library is called **NOT**. The Graphic Editor file for the 4-bit adder/subtractor system should resemble Figure 10-26.

2. The UP board can now be programmed. Lab 1 describes all the steps that are required to program a UP board with the arithmetic system. Use the Lab Work Procedure in Lab 1 as a guide, and complete steps 13 through 17. Use a pushbutton for the **AddSub** control input. The NOT gate will ensure that the system will add when the pushbutton is released, and it will subtract when the pushbutton is held down. Ignore any compiler

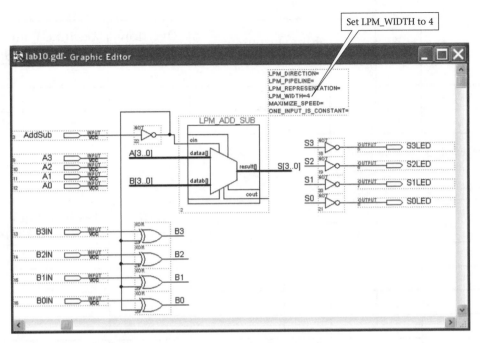

Figure 10-26 Altera "Graphic Editor": 4-Bit Adder/Subtractor System

warning and information messages. You are ready to test the operation of the adder/subtractor system.

LAB EXERCISES AND QUESTIONS

This section contains lab exercises that can be performed on the UP board and questions that can be answered at home. Ask your instructor which exercises to perform and which questions to answer.

Lab Exercises

Exercise 1: Test the 4-bit adder/subtractor system.

The system can be used for 4-bit *unsigned addition* or 4-bit *signed addition/subtraction* (but it cannot do both at the same time).

    a. What is the range of unsigned numbers that can be represented with 4 bits?

    b. What is the range of signed numbers that can be represented with 4 bits?

    c. Use the system for *unsigned* addition and record the results of the operations in a table similar to the one shown in Figure 10-27. The column titled "Overflow" is used to indicate if the math system generates an error. *Overflow* means the answer that is output by the system is incorrect. For example, if A = 10 and B = 9, the sum should be 19. The system cannot generate a 4-bit sum to represent 19. The answer is too big to be fit into a 4-bit number.

    d. Use the system for *signed* addition and subtraction. Record the results of the operations in a table similar to the one shown in Figure 10-28.

    e. The operation 7 + 7 was attempted in step c and step d. Why did it generate an arithmetic overflow in step d but not in step c?

Exercise 2: Use "LPM_ADD_SUB" features to eliminate the XOR gates.

The Altera **LPM_ADD_SUB** symbol includes many optional features. Read the "Help" screen for the **LPM_ADD_SUB** symbol and figure out how to make it add and subtract numbers without the XOR gates. Make the changes, remove the XOR gates, and test the system.

A	+	B	4-Bit Output				Overflow (yes/no)?	Decimal Equivalent (of 4-Bit Output)
3	+	2						
7	+	7						
9	+	9						

Figure 10-27 Exercise 1(c) Unsigned Addition Table

A	+/−	B	4-Bit Output				Overflow (yes/no)?	Decimal Equivalent (of 4-Bit Output)
			Sign	3-Bit Magnitude				
3	+	2						
6	−	2						
3	−	6						
7	+	7						
−1	+	−6						
−1	−	−6						

Figure 10-28 Exercise 1(d) Signed Addition and Subtraction Table

Exercise 3: Add a digital display to the 4-bit adder/subtractor system.

Remove the LED (**S3LED**, **S2LED**, **S1LED**, and **S0LED**) and add the display decoder from Lab 6. The system will display the answer to the math operations on a digital display. Use the system for *signed* addition and subtraction. Record the results of the operations in a table similar to the one shown in Figure 10-29.

A	+/−	B	Digital Display	4-Bit Output				Overflow (yes/no)?
				Sign	3-Bit Magnitude			
6	−	2						
3	−	6						
7	+	7						
−1	+	−6						
−1	−	−6						

Figure 10-29 Exercise 3: Signed Addition and Subtraction Table

Exercise 4: Design a system to display negative answers in readable format.

In Exercise 3, the digital display worked well when the answer to a math operation is positive. For example ,"6 − 2 = 4" is displayed as "4" on the digital display. Unfortunately, "3 − 6 = −3" displays a hexadecimal "D." "D" is the correct 2s complement notation for "−3"; however, it would be more desirable to display "−3." Figure out how to add devices to the system in order to display both positive and negative answers in readable format. Positive answers will continue to be displayed properly and negative answers will also be displayed properly, with a leading minus sign. The minus sign can be displayed by lighting the "g" segment of the second digital display.

Exercise 5: Design a system to display E for "ERROR" whenever arithmetic overflow occurs.

In Exercise 4, the digital display worked well when the answer to a math operation was both positive and negative. Unfortunately, the system does not indicate if an arithmetic overflow has occurred. Figure out how to add devices to the system in order to display both positive and negative results in readable format, along with an error message when arithmetic overflow occurs. When arithmetic overflow occurs display an "E." You can build your own overflow detection system (refer to Introductory Information) or you can use the overflow feature built into the **LPM_ADD_SUB** symbol.

Exercise 6: Use the Altera simulator to test the operation of a 5-bit adder system.

(This exercise can be done at home.)

Use the MAX+PLUS II "Graphic Editor" to draw a diagram for a **5-bit adder system**. Use the Lab Work Procedure in Lab 1 as a guide, and complete steps 1 through 12. Use the project name **lab10b** and save the file in folder **labs**. The adder symbol in the "mega_lpm" library is called **LPM_ADD_SUB**. The Graphic Editor file for the 5-bit adder system should resemble Figure 10-30.

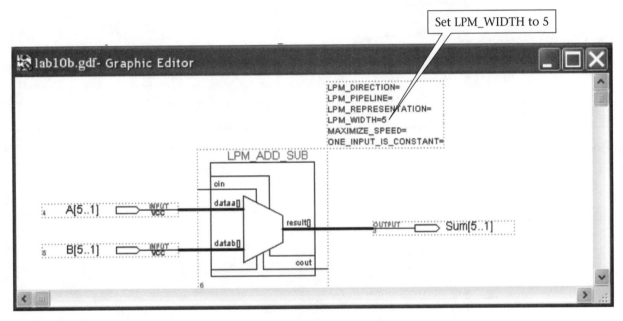

Figure 10-30 Altera "Graphic Editor": 5-Bit Adder System for Simulation Exercise

Use Appendix F as a guide to run a simulation of the 5-bit adder system you have just created. You will be able to use the "Insert node" window to immediately enter a "Group node" for inputs "A," "B," and output "Sum." The "Waveform Editor" window for the simulation should resemble Figure 10-31.

- Set the **End time** to 1.0us. Select **View** and click on **Fit in Window**.
- Place the selection pointer over the node named **A[5..1]**. Right click then select **Overwrite** then **Count Value ...** to open the "Overwrite Count Value" window. Change

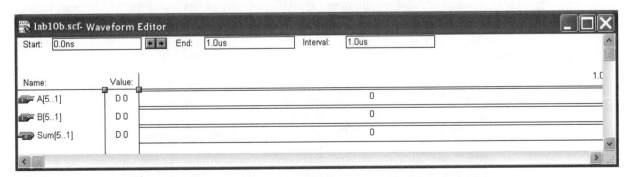

Figure 10-31 Waveform Editor Window: 5-Bit Adder System

the **Count Every** entry box to **50.0ns**. If the "Count Every" entry box is grayed out (not selectable), then you should exit the "Overwrite Count Value" window, select the **Options** menu, and remove the check mark beside the menu item "Snap to Grid." Return to the "Overwrite Count Value" window and you will be able to set the count period. Press the "OK" button, and the waveform for node "A[5..1]" will appear in the "Waveform Editor" window.

- Place the selection pointer over the node named **B[5..1]**. Right click then select **Overwrite** then **Count Value ...** to open the "Overwrite Count Value" window. Change the **Count Every** entry box to **100.0ns**. Press the "OK" button, and the waveform for node "B[5..1]" will appear in the "Waveform Editor" window. The "Waveform Editor" window for the simulation should resemble Figure 10-32. Run the simulation and explain the output response you observe at "Sum[5..1]".

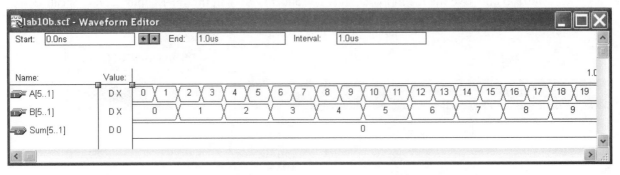

Figure 10-32 Waveform Editor Window: 5-Bit Adder System with Waveforms Defined

Lab Questions

1. Convert the following numbers to 8-bit 2CN:
   a. –32
   b. +89
   c. –87
   d. –128
2. Convert the following 8-bit 2CN numbers to decimal:
   a. 11001011
   b. 01110110
   c. 10101011
   d. 10000110
3. Study the 5-bit combination adder/subtractor system shown in Figure 10-33.
   The system has been constructed and a technician has placed the system in the subtract mode with "A = –12" and "B = +4."
   a. Show the logic "1" and "0" levels inside each box for "A," "B," "S," and "Cout" and at the output of all XOR gates.
   b. Check the binary answer at output "S." Does it equal "–16"?
   c. What is the decimal answer to the subtraction process if the connection from the output of XOR gate number "2" to the "B2" input has a fault on it? The fault causes input "B2" to be a constant "logic 0."

Conclusions

- Adder systems are made up of individual elements called full adders.
- 2s complement notation allows an adder system to do subtraction.
- The 2s complement conversion procedure can be used to convert a positive 2CN number to a negative 2CN number. It can also be used to convert a negative 2CN number to a positive 2CN number.
- Arithmetic overflow is an error that occurs if two numbers of the same sign are added together and generate an answer that exceeds the system's valid range.

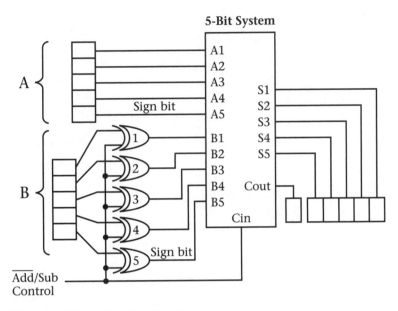

Figure 10-33 Adder/Subtractor System for Question 3

- The total number of bits allocated to the system determines the valid range of the system.
- Arithmetic overflow never occurs if two numbers of opposite signs are added together.
- An arithmetic overflow detection system compares the sign bit of the two numbers being added together to the sign bit of the answer.
- The largest negative 2CN number always has the MSB = 1 and the magnitude bits are all 0s.
- The Altera "LPM_ADD_SUB" symbol has many features that can be used to create combination adder/subtractor systems.

Memory System Fundamentals

Equipment   Altera UP board
Book CD-ROM
Spool of 24 AWG wire
Wire cutters
Blank floppy disk

Objectives   Upon completion of this lab you will be able to:

- Design ROM and RAM systems.
- Store binary data into a RAM.
- Design a ROM light sequencer.

INTRODUCTORY INFORMATION

Memory System Terminology

A memory system is a binary number storage system. Binary numbers (or data) can be stored into the system and later retrieved. The term **writing to memory** is used to describe the process of storing a number into memory. The term **reading** is used to describe the process of retrieving a binary number from memory. A **storage capacity** equation is used to describe how many different numbers the memory system can store. Figure 11-1 shows a memory system capable of storing 16 different 4-bit numbers. It has a storage capacity equation **16×4** (pronounced "16 by 4"). The memory system uses 16 4-bit data registers to store the data.

A **RAM** memory system is a temporary data storage system. A **ROM** memory system is a permanent data storage system. An analogy comparing a school blackboard and a school textbook to RAM and ROM can be used to explain the differences between RAM and ROM. At school the instructor places data on the blackboard (whiteboard or overhead) for a short period

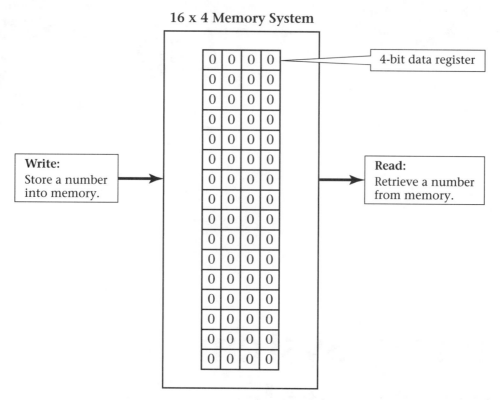

Figure 11-1 A 16×4 Memory System

of time. It can be altered, erased, or changed as needed throughout the presentation. The blackboard is like RAM. A school textbook is like ROM. It stores data permanently and its data can be accessed at any time but it cannot be altered.

The letters *RAM* represent the words **random access memory**. A quick look back at the evolution of memory systems explains the use of the words "random access." In the early days, data was stored on magnetic tape (think of music audiotape). Tape was useful because you could store and retrieve data and later rerecord over the data when you wanted to make changes. Tape systems, however, were sequential access memory systems. To access data at the end of the tape, you would need to fast forward the tape to the data. When semiconductor memory evolved, it created a memory system in which the access time was the same for all data stored in the system. You could randomly access any data and not need to wait for the system to fast forward or rewind.

The letters *ROM* represent the words **read-only memory**. A ROM system is programmed (written) with data once and its data can be read over and over again. In the school textbook analogy, the writing process occurs once by the publisher's printing press. Figure 11-2 shows the differences between RAM and ROM.

Accessing data in memory involves sending a binary number to the memory system that will identify the location that you are trying to access. This number is called an **address**. Once the memory system receives an address, it knows which piece of data you are trying to access. The term *address* is used because it is similar to the number placed on a house (street address) that is used to identify its location in the neighborhood. Figure 11-3 shows how you would read the data stored at address 6 in a 16×4 ROM.

The evolution of memory technology gave birth to many acronyms. The term **nibble** is used to signify a group of 4 bits. The term **byte** is used to signify a group of 8 bits. The term **1 K** is used to round down and signify the number **1024**. To understand the K acronym, look at a 16×4 memory system. A 4-bit address (0 to 15) is required to access data in a 16×4 memory system. A 5-bit address (0 to 31) is required to access data in a 32×4 memory system. A 10-bit address (0 to 1023) is required to access data in a 1024×4 memory system. A 1024×4 memory system is also called a 1 K-nibble memory system. Thus, 1 K = 2^{10} = 1024. *K* is a convenient way of rounding down the number 1024. Here is another example: A **2 K-byte** memory system can store 2048 8-bit numbers. The term **1 M** is used to round down and signify **1 Mega** (1 M = 2^{20} = 1,048,576).

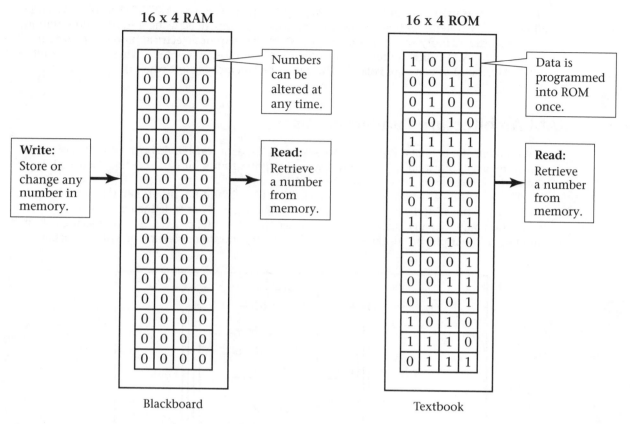

Figure 11-2 RAM and ROM Memory System

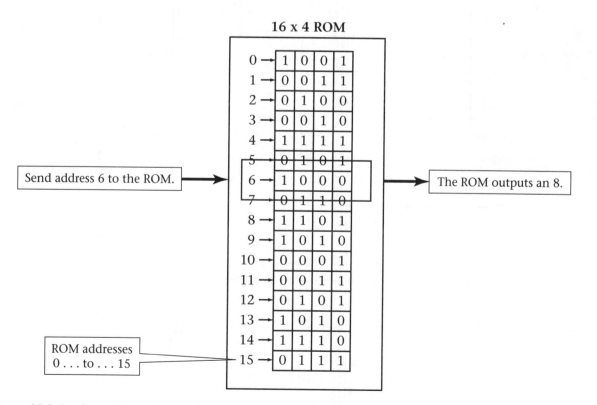

Figure 11-3 Reading address 6 from a 16×4 ROM memory system

The term **1 G** is used to round down and signify **1 Giga** (1 G = 2^{30} = 1,073,741,824). *K*, *M*, and *G* allow memory systems with large storage capacities to be described using a short hand notation.

A **volatile** memory system is a memory system that requires electrical power to maintain data in memory. A **nonvolatile** memory system does not lose its contents when power is shut down. Most RAM systems are volatile. The content of memory is lost when the power is shut down.

RAM Architecture Fundamentals

The fundamental core of a RAM memory system is the data register. A single 4-bit data register can store one 4-bit number. A 4×4 RAM system requires four 4-bit data registers. Figure 11-4 shows the fundamental structure of 4×4 RAM.

The four internal 4-bit registers are labeled R0, R1, R2, and R3. Inputs D0, D1, D2, D3 are connected to the 4-bit data registers in a **bus** configuration. A *bus* is defined as a shared connection among many devices. It is easy to see that the write process involves placing a 4-bit number on the **input bus** (inputs D0, D1, D2, and D3) and then asserting one of the enables

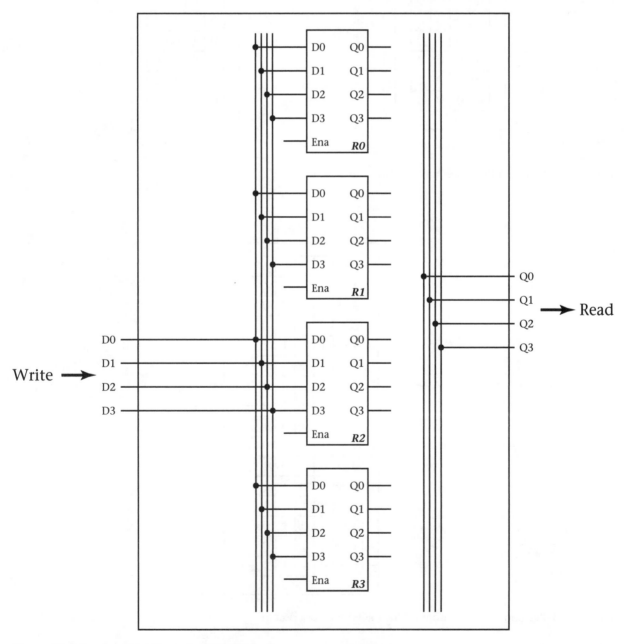

Figure 11-4 Fundamental Core of a 4×4 RAM

(Ena.). To read a number, it would seem logical to connect all four Q0s together, all four Q1s together, all four Q2s together, and all four Q3s together. This would create an **output bus**. The problem is, unlike inputs that are passive, each individual Q output generates an active voltage. Many ICs use 5 volts for logic 1 and 0 volt (or ground) for logic 0. For instance, connecting all four Q0s together to create an output bus would cause a short circuit if one of the Q0 bits is at 0 and another Q0 is at 1. A short circuit theoretically causes an infinite amount of current to flow and would damage the bus. This type of fault has been named **bus contention**. To create an output bus that does not cause "bus contention," you must use a device called a **tri-state buffer**. Tri-state buffers will be studied separately and then they will be used to complete the output bus of the RAM memory system.

Tri-State Buffer Theory

A **buffer** is a device that outputs the same logic level that it receives at the input. Refer to Figure 11-5. At first glance, a buffer seems no more useful than a wire that conducts in one direction from input toward output. Buffers, however, are often used to *isolate* voltage sources of two sections of a system. The input side of a buffer can be connected to one system; meanwhile, the output side can be connected to a second system. The buffer electrically separates the two systems. Buffers can also be used as a **driver** to the second system. A *driver* is a buffer that can handle heavy electrical loads at the output. The first system can process data at low voltage and low current levels and then drive a second system with heavy electrical loads.

Figure 11-5 Buffer

A **tri-state buffer** is a device that has three operating modes. They are 1, 0, and high impedance (HiZ). Refer to Figure 11-6.

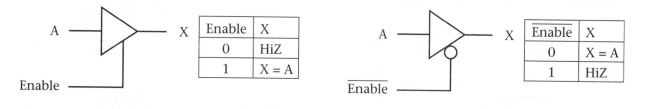

Active High Enable   **Active Low Enable**

Figure 11-6 Tri-State Buffer

Some tri-state buffers have an active high enable input, whereas others have an active low enable input. When the enable is asserted, the buffer operates with a normal logic level 1 or 0 response at the output. When the enable is not asserted, the output goes into a third mode of operation called **high impedance (HiZ)**. High impedance is an output mode that totally disconnects the output from logic 1 and logic 0. "X" is like an output without any power or "X" is like an open circuit. To understand "HiZ" it is useful to study the internal transistor architecture of a digital IC. Figure 11-7 reviews the basic operating principles of an **N-channel MOSFET**.

As seen in Figure 11-7 transistors are used as switches. A high on the gate input closes a switch between the drain and source. A low opens the switch. Figure 11-8 shows the transistor architecture at the output of a tri-state buffer. The architecture is called **CMOS**.

Figure 11-9 shows how the CMOS transistor architecture outputs a "0" and a "1" at "X."

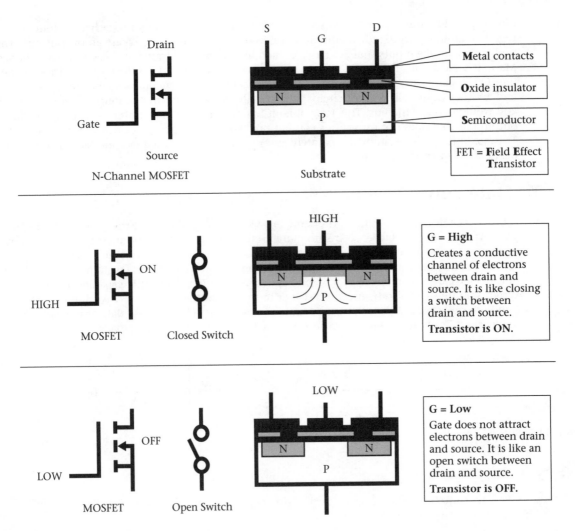

Figure 11-7 N-Channel MOSFET Principles

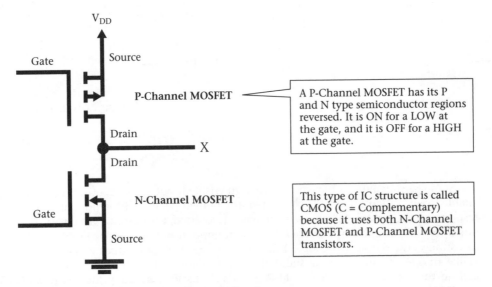

Figure 11-8 CMOS transistor architecture: Tri-State Buffer Output

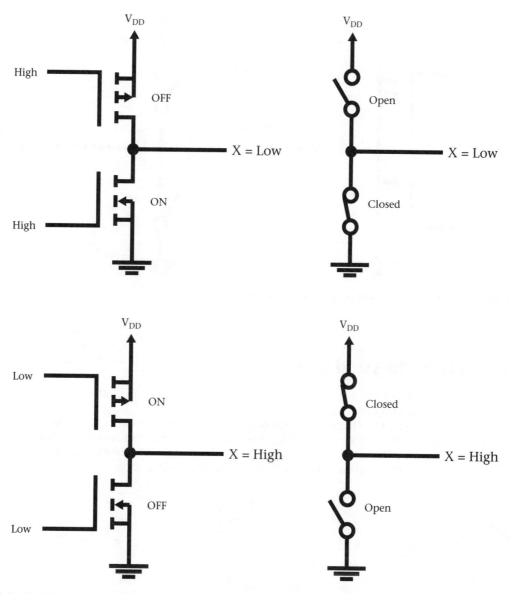

Figure 11-9 CMOS transistor architecture: Tri-state buffer outputs "0" And "1"

Figure 11-9 shows that when both FET gates are high, the N-channel FET turns on and connects "X" to low (0). It also shows that when both FET gates are low, the P-channel FET turns on and connects "X" to high (1). The FETs connect output "X" to either a high or a low. Figure 11-10 shows how a tri-state buffer outputs HiZ.

Figure 11-10 shows that when the gate is low, the N-channel FET turns off and disconnects "X" from low (0). It also shows that when the gate is high, the P-channel FET turns off and disconnects "X" from high (1). A "HiZ" at the output completely disconnects output "X" from a high and a low because both FETs are off. It is like having open switches to Vdd and ground.

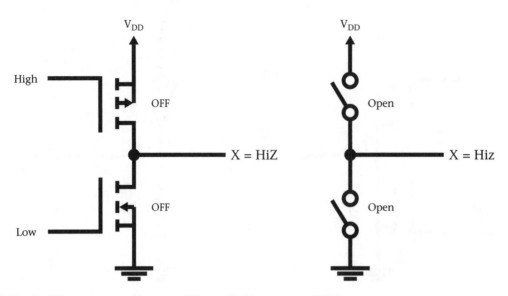

Figure 11-10 CMOS transistor architecture: Tri-state buffer outputs "HiZ"

Tri-State Buffers and RAM

Figure 11-11 shows the RAM memory system with tri-state buffers. It also shows how you would access the data in register R2.

At output Q0, Q1, Q2, and Q3, bus contention has been eliminated. The tri-state buffers for R0, R1, and R3 are in "HiZ." "HiZ" disconnects the highs and lows from those data registers and only allows the data from R2 to be read. On the input side, R2 is enabled while R0, R1, and R3 are in the hold mode. Data can be written only to data register R2.

4X4 RAM

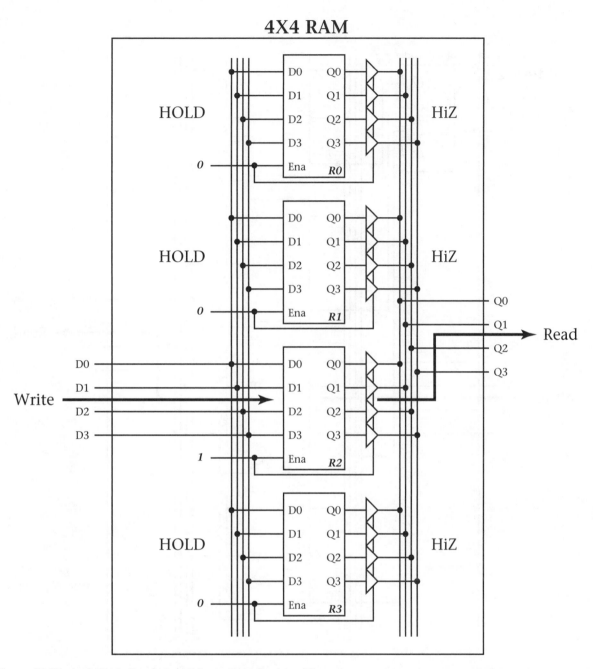

Figure 11-11 4×4 RAM: Read and Write to Data Register R2

RAM Addressing

Accessing data in memory involves sending a binary number to the memory system that will identify the location that you are trying to access. This number is called an **address**. Once the memory system receives an address, it knows which piece of data you are trying to access. Figure 11-12 shows how a "1 of 4" decoder is used to select register 2. Each register (R0, R1, R2, and R3) has its own unique address. The decoder also ensures that bus contention cannot occur.

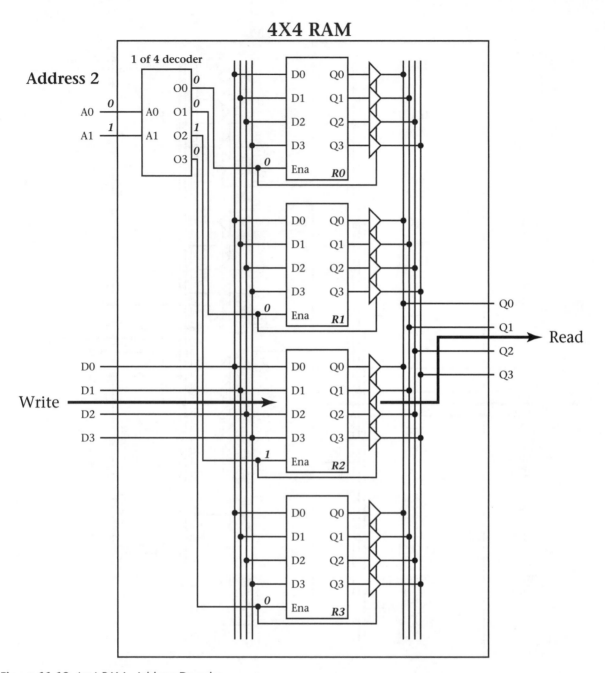

Figure 11-12 4×4 RAM: Address Decoder

Separating the "Read" and "Write" Operations

Logic must be added to the system in order to separate the function of reading and writing. The current RAM system uses the address decoder to select a register. The register is enabled and its data can be read. The enabled register also stores data from D0, D1, D2, and D3 at the very same time. This means you cannot read the memory system without having to write to it at the same time. Figure 11-13 shows how AND gates can be used to create a system that can separate the read and write operations.

The write enable control input can be used to control the flow of data into or out of the memory device: **WE = 1** for write operations and **WE = 0** for read operations.

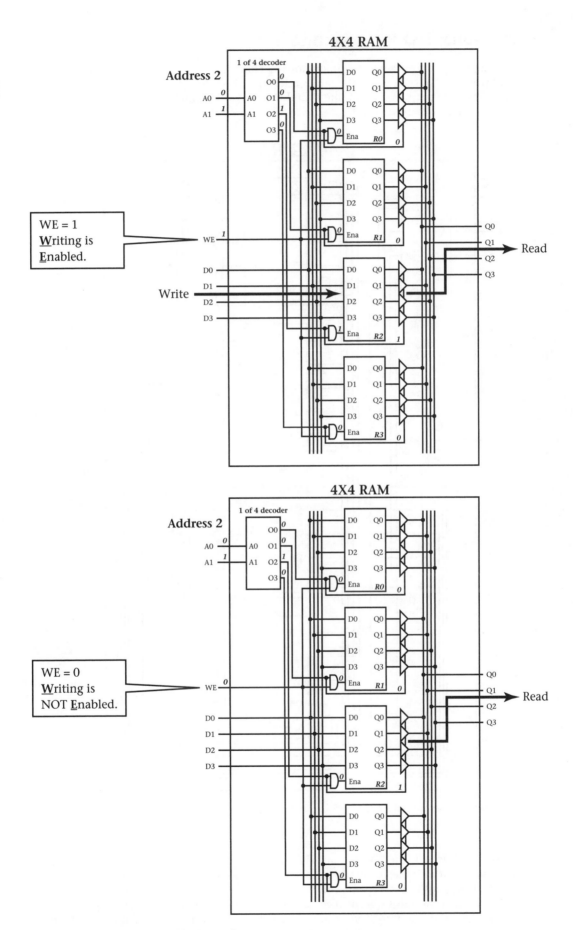

Figure 11-13 4×4 RAM: Write Enable Control

Bidirectional I/O RAM Architecture

Most digital systems use separate operations to read and write RAM. The RAM never needs to be "read from" or "written to" at the same time. For this reason there is no need to provide separate D inputs and Q outputs. Combining D and Q reduces the number of inputs and outputs. It creates a bidirectional I/O RAM. Refer to Figure 11-14.

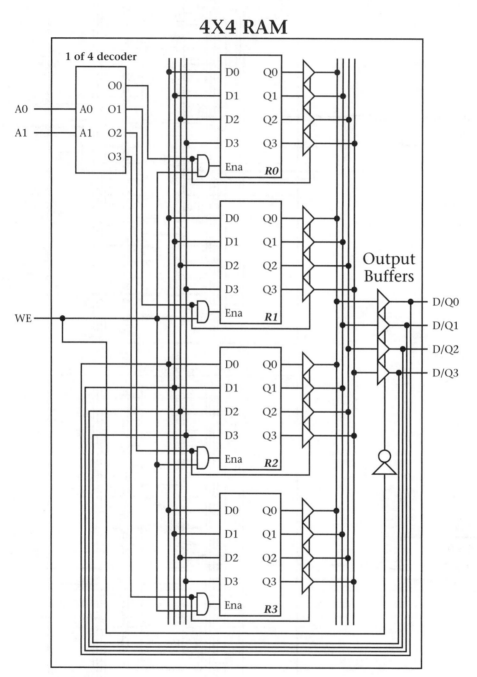

4X4 RAM

Figure 11-14 4×4 RAM: Bidirectional IO

When "WE" = 1, the "Output Buffers" are in HiZ and the D/Q lines can safely write data into the registers without bus contention. An active voltage source (like a switch) is required to drive the "D/Q" lines. When "WE" = 0, the "Output Buffers" will allow data to be read from the registers. Data is output from the registers to the "D/Q" lines. The active voltage source, which was used to write data into memory, must be removed in order to prevent bus contention. A passive device like an LED can be used to view the contents of memory.

Some RAM systems have two additional control inputs named "OE" and "CS." Figure 11-15 shows a bidirectional RAM with "OE" and "CS" control inputs. **OE** is called **output enable**. "OE" can be used to enable the "Output Buffers" to be used to write and read data. "OE" can also be used to completely disable the "Output Buffers" and keep the "D/Q" lines continuously in "HiZ." When "OE" = 1, the "Output Buffers" are enabled and are, in turn, controlled by "WE." When "OE" = 0, the "Output Buffers" are *not* enabled and can no longer be controlled by "WE." They remain continuously in "HiZ" regardless of the condition at "WE."

CS is called **chip select**. Unlike "OE," which only controls the status of the "Output Buffers," "CS" can be used to enable or totally disable the entire RAM semiconductor (chip). When "CS = 1," the "1 of 4" decoder is enabled and the RAM operates normally. Write and read operations can be performed. When "CS = 0," the "1 of 4" decoder is disabled and all decoder outputs are "0." With the decoder outputs at "0," all internal registers (R0, R1, R2, R3) will have their local buffers in "HiZ" and their "Ena" inputs disabled. This totally shuts down any write and read operations.

4X4 RAM

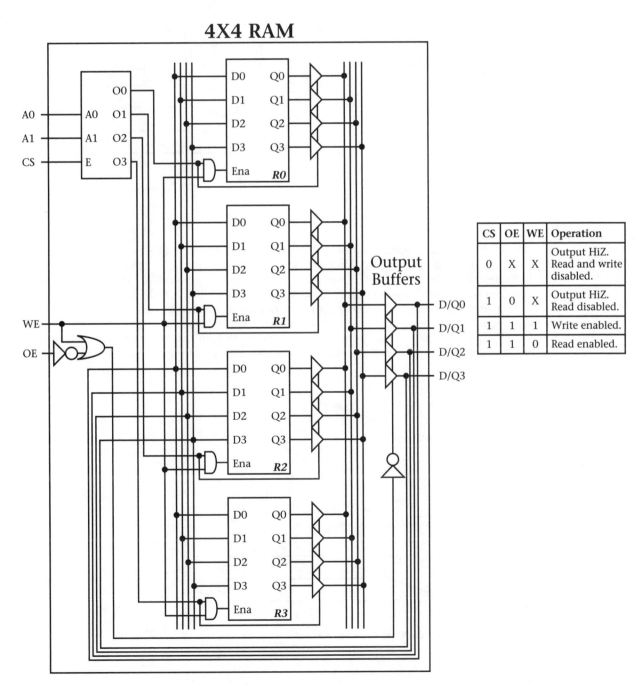

CS	OE	WE	Operation
0	X	X	Output HiZ. Read and write disabled.
1	0	X	Output HiZ. Read disabled.
1	1	1	Write enabled.
1	1	0	Read enabled.

Figure 11-15 4×4 RAM: Bidirectional RAM with "OE" and "CS"

ROM Architecture Fundamentals

A ROM stores predetermined binary data. Once the data is stored (or programmed) then it can only be read. The fundamental core of a RAM is a data register. The fundamental core of a ROM is a data register without data inputs. This sounds strange, but if you think about it for a few seconds, it makes sense. A data register without D inputs can only be read at outputs Q. There are many techniques that can be used to program a ROM. A look back in time to an era when ROMs were available in a fuse programmable package is a good starting point. Figure 11-16 shows the fuse structure of a ROM.

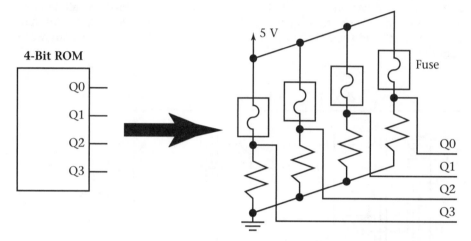

Figure 11-16 Fundamental Core of a 4-Bit ROM

The 3-D fuse-resistor diagram shows how each bit in the memory cell structure is a fuse to 5 volts (high) and a resistor to ground (low). Leaving the fuse intact stores a "1" in the memory cell. Burning out the fuse stores a "0." When a fuse is burned out, the cell is connected to ground through the resistor. A ROM programmer is a device that takes a blank ROM (all fuses intact) and uses a programming voltage (high voltage) to burn out fuses, which correspond to a "0" in memory. Figure 11-17 shows how the ROM programmer would burn out fuses to store the number "9" in ROM. Because you cannot write to a ROM, the internal architecture is less complicated. Figure 11-18 shows the architecture of a 4×4 ROM with an active low chip select.

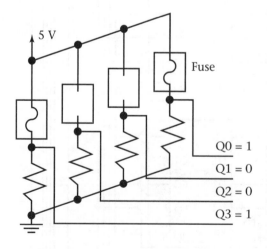

Figure 11-17 4-Bit ROM Programmed with the Number "9"

4X4 ROM

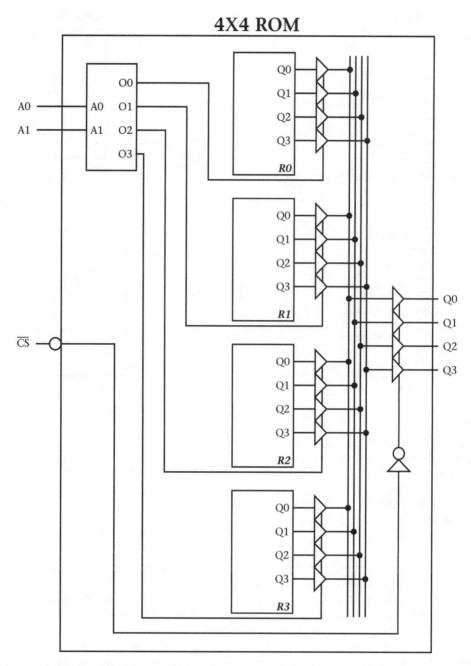

Figure 11-18 4×4 ROM with an Active Low CS

RAM and ROM Schematic Symbols

The schematic symbol for a 4×4 RAM and a 4×4 ROM is shown in Figure 11-19. The input/output connections to all memory systems can be broken down to three groups: **data bus**, which provides access to the content of the memory cells; **address bus**, which selects which memory cells are transferred to and from the memory system; and **control bus**, which ontrols the direction of data transfer and can be used to disable the memory system. Figure 11-20 shows the schematic diagram of an 8×4 ROM and a 4×8 ROM.

The 8×4 ROM stores eight different 4-bit numbers (nibbles). The 4×8 ROM stores four different 8-bit numbers (bytes). The "First Number" in the storage capacity equation can be used to determine the number of address bus inputs. Solve for "n" in the equation:

First Number = 2^n

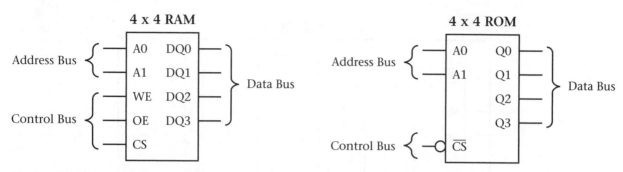

Figure 11-19 4×4 RAM and 4×4 ROM Schematic Symbols

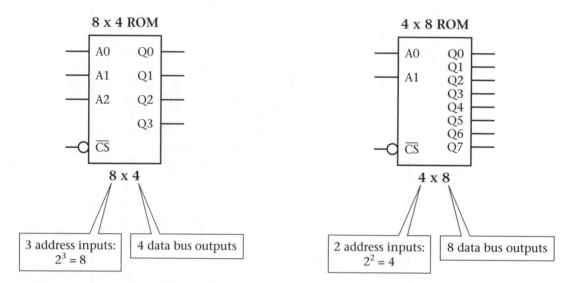

Figure 11-20 8×4 ROM and 4×8 ROM Schematic Symbols

The number of data bus outputs is equal to the second number in the storage capacity equation. The **bit storage capacity** of a memory system is obtained by multiplying both numbers of the storage capacity equation. An 8×4 ROM can store 8 × 4 = 32 bits of data. A 4×8 ROM can store 4 × 8 = 32 bits of data. Both ROMs store the same number of total bits (32) but they do so very differently.

ROM Light Sequencer System and Memory Map Diagrams

A memory map diagram is used to represent *what* data is stored inside a memory system. Figure 11-21 shows a ROM light sequencer system and a binary/hex memory map diagram.

The binary memory map diagram is useful because it shows the actual data that has been programmed into the ROM. For example, at address "0111," the binary number "01000010" is stored. Remember a blank ROM has all of its fuses intact and the entire memory would be filled with 1s. The binary memory map diagram is filled with a mix of 1s and 0s, indicating that some fuses are burned and others are intact. The HEX memory map diagram conveys the same information, with all binary numbers converted to HEX. Look at the same example: At address 7 (0111), the HEX number stored is 42 (4 = 0100 and 2 = 0010). Binary memory map diagrams are rarely used because the binary numbers are typically too large to work with. HEX numbers are compact and easier to work with. It is easier to remember that address 7 = 42 than to try and remember that address 0111 = 01000010.

To convey the importance of a memory map diagram, let's study the operation of the **light sequencer** system. A light sequencer system has a ROM controlling a set of LEDs. The data stored inside the ROM determines the on and off status of the LEDs. A designer decides how the lights

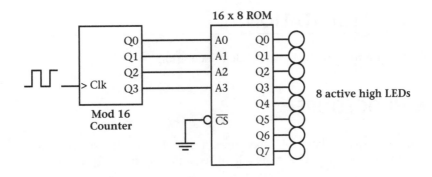

A_3 A_2 A_1 A_0	$Q_7 - Q_0$
0	81
1	42
2	24
3	18
4	00
5	18
6	24
7	42
8	81
9	42
A	24
B	18
C	00
D	18
E	24
F	42

A_3	A_2	A_1	A_0	Q_7	Q_6	Q_5	Q_4	Q_3	Q_2	Q_1	Q_0
0	0	0	0	1	0	0	0	0	0	0	1
0	0	0	1	0	1	0	0	0	0	1	0
0	0	1	0	0	0	1	0	0	1	0	0
0	0	1	1	0	0	0	1	1	0	0	0
0	1	0	0	0	0	0	0	0	0	0	0
0	1	0	1	0	0	0	1	1	0	0	0
0	1	1	0	0	0	1	0	0	1	0	0
0	1	1	1	0	1	0	0	0	0	1	0
1	0	0	0	1	0	0	0	0	0	0	1
1	0	0	1	0	1	0	0	0	0	1	0
1	0	1	0	0	0	1	0	0	1	0	0
1	0	1	1	0	0	0	1	1	0	0	0
1	1	0	0	0	0	0	0	0	0	0	0
1	1	0	1	0	0	0	1	1	0	0	0
1	1	1	0	0	0	1	0	0	1	0	0
1	1	1	1	0	1	0	0	0	0	1	0

Hex memory map diagram:
(Mostly used).

Binary memory map diagram:
(Rarely used).

Figure 11-21 16×8 ROM Light Sequencer System

will be sequenced and programs the ROM. The counter is used to cycle the address bus. Each count state generates an address and the ROM outputs an 8-bit number to the LEDs. The clock rate of the counter determines how long the LEDs will display the data from ROM before new data is sent to the LEDs. The clock rate, thus, controls the light sequencing speed. If you look at the pattern in the binary memory map diagram, you will see that two lit LEDs appear at the outer edge and move toward the middle and back out again. The cycle is repeated twice. The advantage of a ROM-based system is in the flexibility to make changes. Reprogramming a ROM and inserting it into the system can easily change the light sequence. There is no need to change any other system components or connections.

POWERPOINT PRESENTATION

Use PowerPoint to view the LAB 11 slide presentation.

LAB WORK PROCEDURE

There are two separate lab work procedures. In part 1, you will use a RAM to store today's date. In part 2, you will use a ROM to create a light sequencer system.

Lab Work Procedure Part 1: Store Today's Date into an 8×4 RAM

Although some RAM configurations work with the MAX IC, it is recommended that you use the FLEX IC. You will build an 8×4 RAM system and use it to manually store binary data that represents today's date. Manually operating the RAM will teach you the sequence a computer must use to store and retrieve data.

1$_{FLEX}$. You will draw a diagram for an 8×4 RAM system. Use the Lab Work Procedure in Lab 1 as a guide, and complete steps 1 through 12. Begin by creating a folder (directory) on your blank disk named **Labs**. Use the project name **Lab11.gdf**. The RAM symbol in the "mega_lpm" library is called **LPM_RAM_DQ**. The Graphic Editor file for the 4-bit 8×4 RAM system should resemble Figure 11-22.

2$_{FLEX}$. The UP board can now be programmed. Lab 1 describes all the steps that are required to program a UP board with the 8×4 RAM system. Use the Lab Work Procedure in Lab 1 as a guide, and complete steps 13 through 17. Ignore any compiler warning and information messages. Assign FLEX switches to the address inputs and the data inputs. Assign a FLEX pushbutton to the WE input. The pushbutton is 1 when it is not pressed and changes to 0 when it is pressed and held down. Assign FLEX LEDs to outputs **Qout0**, **Qout1**, **Qout2**, and **Qout3**. You are ready to test the operation of the 8×4 RAM system.

3$_{FLEX}$. The current date will be stored in RAM using the following eight-digit format: **MM DD YYYY**. Fill in the "Data In" column of the binary memory map diagram shown in Figure 11-23.

4$_{FLEX}$. Use the address, data in, and write enable switches to write the current date into the RAM. Use the memory map diagram generated in step 3 as a guide.

5$_{FLEX}$. Use the address switches and data out LEDs to read the current date from the RAM. Repeat step 4 to correct any errors. Fill in the "Data Out" column of the binary memory map diagram shown in Figure 11-23.

6$_{FLEX}$. Add a digital display to the 8×4 RAM system. Remove the LEDS (Qout0, Qout1, Qout2, and Qout3) and add the display decoder. Use the display decoder that is part of the Altera symbol library. The display decoder symbol in the "mf" library is called **7447**. The system will display the eight digits of the current date on a digital display. Repeat steps 3, 4, and 5.

7$_{FLEX}$. Answer the following questions:
a. Explain how the "WE" pushbutton is used to "write" and "read" the RAM.
b. What could happen if you press the "WE" pushbutton when reading back the date from RAM?

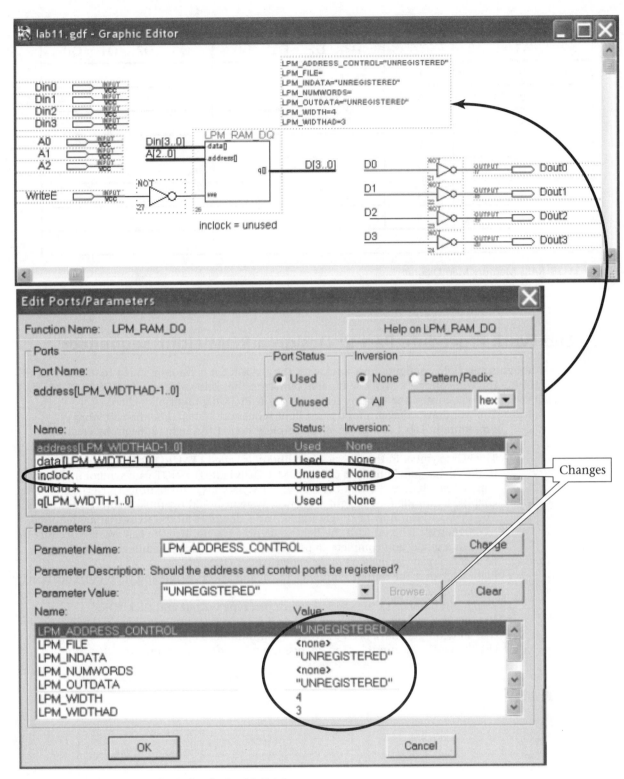

Figure 11-22 Altera "Graphic Editor": 8×4 RAM System

Address			Date	Data In				Data Out			
A2	A1	A0		D3	D2	D1	D0	Q3	Q2	Q1	Q0
0	0	0	M								
0	0	1	M								
0	1	0	D								
0	1	1	D								
1	0	0	Y								
1	0	1	Y								
1	1	0	Y								
1	1	1	Y								

Figure 11-23 RAM Memory Map Diagram

Lab Work Procedure Part 2: Design a ROM Light Sequencer

Although some ROM configurations work with the MAX IC, it is recommended that you use the FLEX IC. You will connect a MOD 32 binary counter to the address inputs of a 32×16 ROM. The ROM outputs will control the lighting sequence of the FLEX digital display LEDs.

1$_{FLEX}$. You will draw a diagram for a ROM light sequencer system. Use the Lab Work Procedure in Lab 1 as a guide, and complete steps 1 through 12. Begin by creating a folder (directory) on your blank disk named **Labs**. Use the project name **Lab11B.gdf**. The ROM symbol in the "mega_lpm" library is called **LPM_ROM**. The counter symbol in the "mega_lpm" library is called **LPM_COUNTER**. The NOT gate symbol in the "mega_lpm" library is called **LPM_INV**. "LPM_INV" is equivalent to 16 NOT gates. The Graphic Editor file for the light sequencer system should resemble Figure 11-24.

2$_{FLEX}$. The UP board can now be programmed. Lab 1 describes all the steps that are required to program a UP board with the light sequencer system. Use the Lab Work Procedure in Lab 1 as a guide, and complete steps 13 through 17. New procedures are listed next.

To assign the digital display pin numbers to the output symbol Q[15..0]:

a. Right click over the output symbol **Q[15..0]** and select **Assign** and then **Pin/Location/Chip**.

b. Assign pin number **17** to "Q[15..0]" with **pin type output** and click "OK."

c. Use the "Assign" menu to return to the "Pin/Location/Chip" window. You will now see that all Q outputs have been assigned to pin number 17.

d. Use the table in Figure 11-25 to change the "Q" output pin numbers.

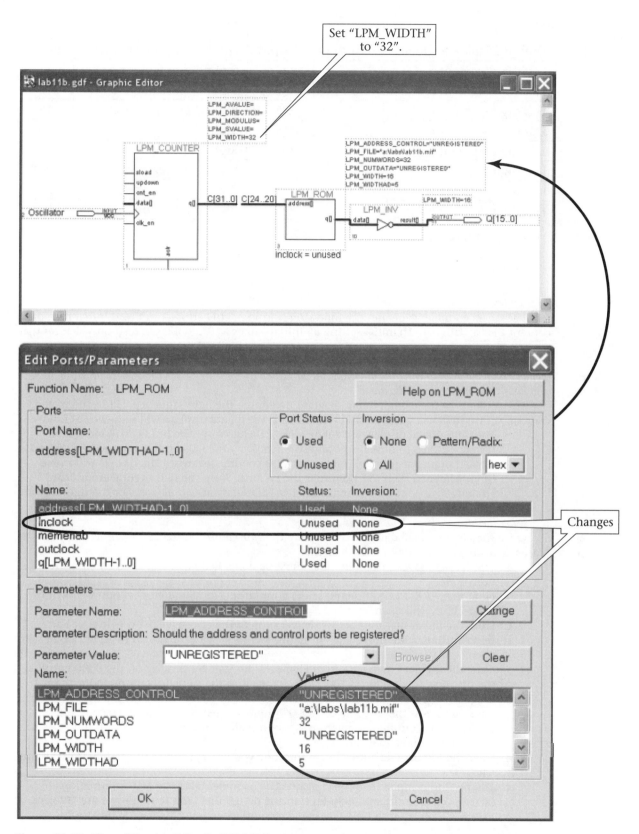

Figure 11-24 Altera "Graphic Editor": ROM Light Sequencer System

ROM Output	Digital Display Segment	Pin #
Q15	Left digit segment *a*	6
Q14	Left digit segment *b*	7
Q13	Left digit segment *c*	8
Q12	Left digit segment *d*	9
Q11	Left digit segment *e*	11
Q10	Left digit segment *f*	12
Q9	Left digit segment *g*	13
Q8	Left digit *decimal point*	14

ROM Output	Digital Display Segment	Pin #
Q7	Right digit segment *a*	17
Q6	Right digit segment *b*	18
Q5	Right digit segment *c*	19
Q4	Right digit segment *d*	20
Q3	Right digit segment *e*	21
Q2	Right digit segment *f*	23
Q1	Right digit segment *g*	24
Q0	Right digit *decimal point*	25

Figure 11-25 FLEX Digital Display Pin Assignments

To initialize the content of ROM:

The Altera compiler generates the following warning: " I/O error—can't read initial memory content file "a:\labs\lab11b.mif"—setting all initial values to 0". ROMs are read-only devices that need to be initialized (or programmed) with data. The initialization comes from the **Memory Initialization File** (MIF) **a:\labs\lab11b.mif**. To create this initialization file, you must *fill in a memory map diagram to enter the data into the MIF file*.

Fill in the memory map diagram:

The contents of the memory map diagram determine the light sequence. Fill in the memory diagram shown in Figure 11-26 so that the LEDs on the digital displays will sequence as follows:

- The right digital display will rotate a lit LED clockwise. The lit LED will move from segment a then to b then to c and so forth. Keep the LED cycling around and around.
- The left digital display will rotate a lit LED counterclockwise. The lit LED will move from segment a then to f then to e and so forth. Keep the LED cycling around and around. Figure out how to make the left LED rotate at *half the speed* of the right digital display LED.

Enter the memory map data into the MIF file:

Once the memory map diagram is complete the MIF can be created. Follow this procedure to place the memory map data into the MIF file:

   a. The "Project Save & Compile" step must be done. Redo this step if you are unsure it has been done.

   b. From the **MAX+PLUS II** menu, select **Simulator**. The "Simulator" window shown in Figure 11-27 will open.

   c. From the menu that runs across the very top of the screen, click on **Initialize** then select **Initialize Memory**. The "Initialize Memory" window shown in Figure 11-28 will open.

   d. Enter the memory map data from Figure 11-26. Start by clicking on the **0000** entry on the first line beside address "00" and change it to the value in your **memory map diagram**. Press the <tab> key and change the next "0000" entry on the right. Work your way across the first line of "0000" numbers then proceed to the second line (address 07). Continue to make changes until all four lines are filled with the ROM data from memory map diagram.

   e. Click the **Export File** button. Save the code you just entered as file **a:\labs\lab11b.mif**. Click "OK."

   f. You need to compile the project again. The compiler takes the MIF information (a:\labs\lab11b.mif) and includes it in the design files. You must always use "Project Save & Compile" each time you make memory changes using the "Simulator."

   g. Configure (program) the FLEX IC and observe the light sequencer.

   h. Calculate how fast each group of LEDs are cycling. In other words, how much time does an LED stay lit before the light sequencer lights a different LED?

Hex Address	ROM Binary Output Q15 . . . . Q8								ROM Binary Output Q7 . . . . Q0								ROM Hexadecimal Output
	a	b	c	d	e	f	g	dp	a	b	c	d	e	f	g	dp	
0																	
1																	
2																	
3																	
4																	
5																	
6																	
7																	
8																	
9																	
A																	
B																	
C																	
D																	
E																	
F																	
10																	
11																	
12																	
13																	
14																	
15																	
16																	
17																	
18																	
19																	
1A																	
1B																	
1C																	
1D																	
1E																	
1F																	

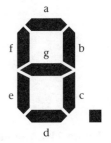

Figure 11-26 ROM Memory Map Diagram

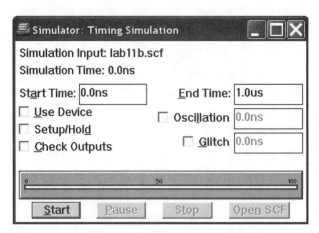

Figure 11-27 Simulator Window

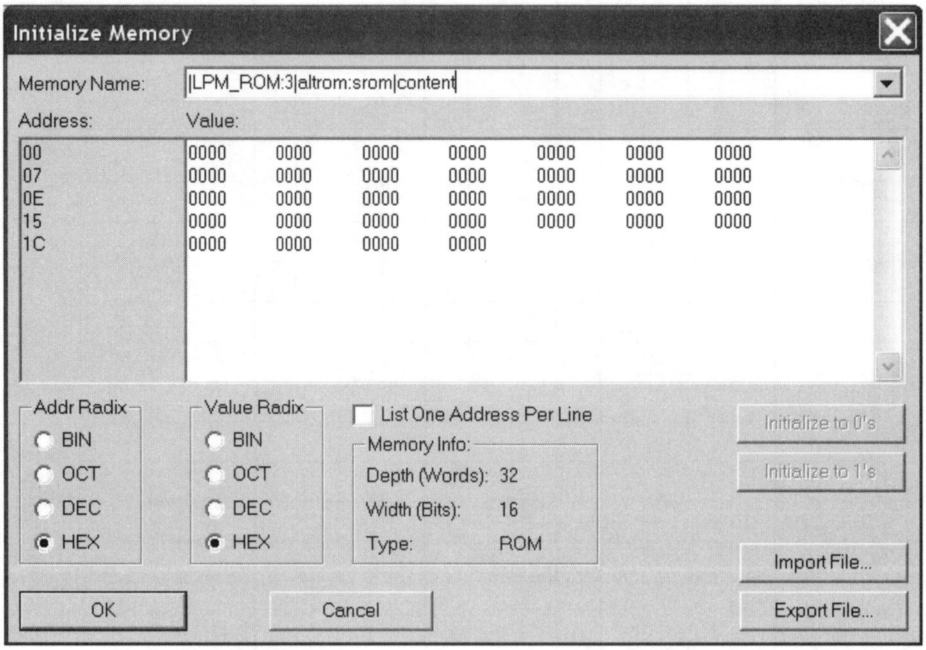

Figure 11-28 Initialize Memory Window

LAB EXERCISES AND QUESTIONS

This section contains lab exercises that can be performed on the UP board and questions that can be answered at home. Ask your instructor which exercises to perform and which questions to answer.

Lab Exercise

Exercise 1: Change the light pattern of the ROM light sequencer system.

In part 2 of this lab you created a ROM light sequencer with rotating lit LED. In this exercise you will change the lighting sequence. Create a new memory map diagram for the ROM light sequencer system. You can make any number of lit LEDs cycle around or blink in different patterns. Be creative. A new memory map diagram is provided in Figure 11-29. Calculate how fast the lit LED is moving around.

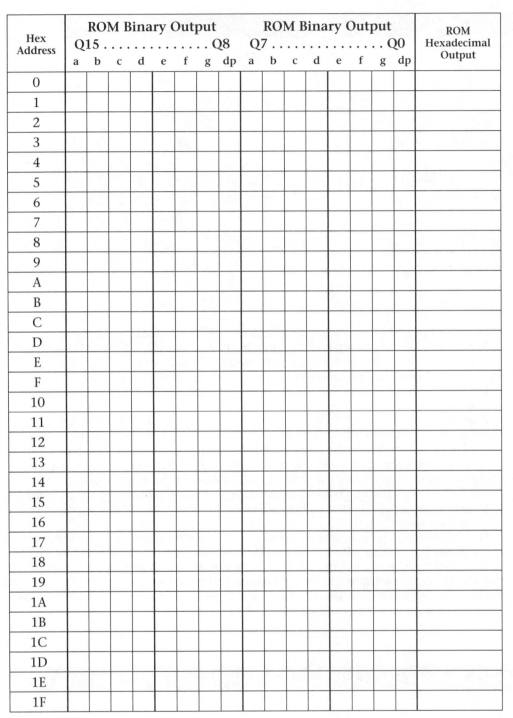

Hex Address	ROM Binary Output Q15 . . . . Q8							ROM Binary Output Q7 . . . . Q0							ROM Hexadecimal Output
	a	b	c	d	e	f	g dp	a	b	c	d	e	f	g	dp	
0																
1																
2																
3																
4																
5																
6																
7																
8																
9																
A																
B																
C																
D																
E																
F																
10																
11																
12																
13																
14																
15																
16																
17																
18																
19																
1A																
1B																
1C																
1D																
1E																
1F																

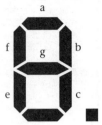

Figure 11-29 ROM Memory Map Diagram

Exercise 2: Use the Altera simulator to test the 8×4 RAM.

(This exercise can be done at home)

In part 1 of this lab you stored the current date in an 8×4 RAM. You will now use Altera simulator to test the operation of the RAM system. Use the MAX+PLUS II "Graphic Editor" to make changes to the lab11.gdf file for the 8x4 RAM system. The changes will simplify the system and allow you to use the simulator to test its operation.

1. Add four NOT gate symbols to the "Dout" outputs. Remove the NOT gate connected to the WE input. It was used to invert the FLEX pushbutton and it is no longer needed. The changes are shown in Figure 11-30. Use Appendix F as a guide to run a simulation

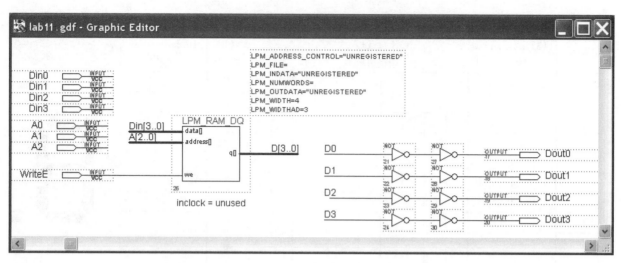

Figure 11-30 Altera "Graphic Editor": 8×4 RAM System Simulation

of the 8×4 RAM system. You will be able to use the "Insert node" window to immediately enter a "Group node" for inputs **A[2..0]** and **Din[3..0]**. The "Waveform Editor" window for the simulation should resemble Figure 11-31.

2. Set the **End time** to 1.0us. Select **View** and click on **Fit in Window**.
3. Place the selection pointer over **WriteE**. Right click and then select **Overwrite** then **High(1)**.
4. Place the selection pointer over the node named **A[2..0]**. Right click then select **Overwrite** then **Count Value ...** to open the "Overwrite Count Value" window. Change the **Count Every** entry box to **120.0ns**. If the "Count Every" entry box is grayed out (not selectable), then you should exit the "Overwrite Count Value" window, select the **Options** menu, and remove the check mark beside the menu item "Snap to Grid." Return to the "Overwrite Count Value" window and you will be able to set the count period. Press the "OK" button and the waveform for node "A[2..0]" will appear in the "Waveform Editor" window.
5. Place the selection pointer over the node named **Din[3..0]**. Right click then select **Overwrite** then **Count Value ...** to open the "Overwrite Count Value" window. Change the **Count Every** entry box to **120.0ns**. Press the "OK" button and the waveform for node "Din[3..0]" will appear in the "Waveform Editor" window.
Group the outputs **Dout0**, **Dout1**, **Dout2**, and **Dout3**. The "Waveform Editor" window for the simulation should resemble Figure 11-32.

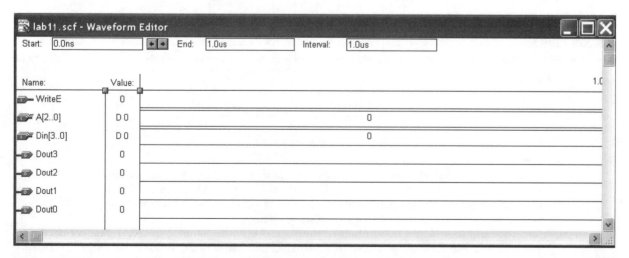

Figure 11-31 Waveform Editor Window: 8×4 RAM System

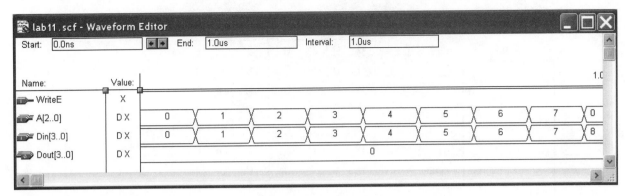

Figure 11-32 Waveform Editor Window: 8×4 RAM System: Input Waveforms Defined

The **Din** waveform represents the data to be stored in RAM. Figure 11-32 shows that the numbers stored in RAM will be the numbers 0 through 7. Change the **Din** waveform to represent the data for **today's date**. Use the following 8-digit format: **MM DD YYYY**. To change **Din**, double-click a "Din" number (**0, 1, ... or ... 7**) then right click and select **Overwrite** then **Group Value ....** The "Overwrite Group Value" window will open. Change the "Group Value" entry box to a digit that represents one of the eight numbers of today's date. Repeat this process to change the remaining seven digits to numbers that represent today's date.

    a. Run the simulation and explain the output response you observe at "Dout[3..0]."

    b. Change **WriteE** to **0** and run the simulation. Explain the output response you observe at "Dout[3..0]."

Lab Questions

1. What is the storage capacity equation for a ROM that has 16 address inputs and 4 data outputs?
2. A RAM has storage capacity equation 8Kx16.
    a. How many data input/output(s) does it have?
    b. How many address bus inputs does it have?
    c. What is the total storage capacity of the RAM in bytes?
3. Define the following four terms: RAM, ROM, volatile, and tri-state buffer.
4. Study the ROM and memory map diagram in Figure 11-33. Determine the HEX value at the data outputs for each of the following input conditions:
    a. A3 A2 A1 A0 = 0111; CS = 1
    b. A3 A2 A1 A0 = 0001; CS = 0
    c. A3 A2 A1 A0 = 1011; CS = 0
    d. A3 A2 A1 A0 = 1111; CS = 0
5. Study the ROM light sequencer system and memory map diagram in Figure 11-34. Fill in the memory map diagram to show how the light sequencer would be programmed to move a lit LED from the far left to the far right and then back from the far right to the far left. It will look like the LED system mounted on the front of the car on the old TV show *Knight Rider*.

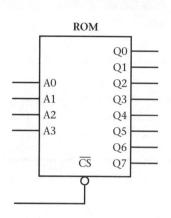

A_3	A_2	A_1	A_0	Q_7	Q_6	Q_5	Q_4	Q_3	Q_2	Q_1	Q_0
0	0	0	0	1	1	0	0	1	0	0	1
0	0	0	1	0	1	1	0	0	0	1	0
0	0	1	0	0	0	1	0	0	1	0	0
0	0	1	1	0	0	1	1	1	0	0	0
0	1	0	0	0	0	0	0	0	0	0	0
0	1	0	1	1	0	0	1	1	0	0	0
0	1	1	0	0	0	1	1	0	1	0	0
0	1	1	1	1	1	0	1	1	1	1	0
1	0	0	0	1	0	0	1	0	0	0	1
1	0	0	1	1	0	1	0	0	0	0	0
1	0	1	0	1	0	1	0	1	0	1	0
1	0	1	1	0	1	0	1	1	1	0	1
1	1	0	0	1	1	1	0	1	0	1	0
1	1	0	1	0	1	0	1	0	1	0	1
1	1	1	0	1	1	1	0	1	1	0	0
1	1	1	1	1	0	1	1	0	0	1	1

Figure 11-33 ROM and Memory Map Diagram for Question 4

Conclusions:

- RAM is used for temporary data storage. It is like a school blackboard.
- ROM is used for permanent data storage. It is like a school textbook.
- A storage capacity equation represents the amount of data a memory system can store.
- Memory devices use an address to identify the location of stored data.
- *Read* means retrieve memory data. *Write* means store memory data.
- HiZ allows many storage cells to share a common set of conductors called a bus.
- Bus contention is a fault that occurs if more than one device uses a bus at the same time.
- Bidirectional I/O RAM uses a single bus to input and output data.
- Memory map diagrams are used to show the data that is stored in memory.

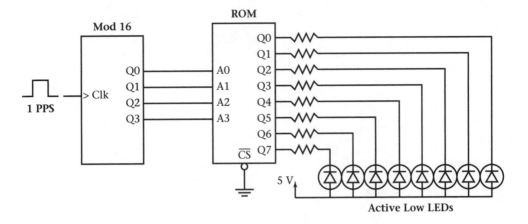

A3 A2 A1 A0 Hex	Q7 – Q0 Binary	Q7 – Q0 Hex
0		
1		
2		
3		
4		
5		
6		
7		
8		
9		
A		
B		
C		
D		
E		
F		

Figure 11-34 ROM Light Sequencer and Memory Map Diagram for Question 5

Liquid Crystal Displays (LCD)

Equipment

Altera UP board

Book CD-ROM

Spool of 24 AWG wire

Wire cutters

Blank floppy disk

Seven segment LCD

Hitachi HD44780 Dot Matrix Display

Objectives

Upon completion of this lab, you should be able to:

* Understand the operation of an LCD.

* Display messages using a two-line 16-character Hitachi HD44780 Dot Matrix Display (DMD).

INTRODUCTORY INFORMATION

In Lab 6 digital displays were introduced as a convenient and user-friendly way of displaying data. Seven-segment digital displays used seven energy-intensive LEDs. Each LED segment required milliamps (10^{-3}) of current. A *l*iquid *c*rystal *d*isplay (LCD) is very energy efficient. An LCD requires only nanoamps (10^{-9}) of current per segment. This energy efficiency is a huge benefit for battery-operated systems like watches and calculators. You will begin by studying seven-segment LCDs and then proceed to two-line 32-character LCD capable of displaying complete messages.

Seven-Segment LCD Construction Diagram

Figure 12-1 shows the construction diagram of a seven-segment LCD. An LCD is made up of **nematic** fluid placed between two glass plates. Transparent conductive segment patterns are etched onto the top glass plate. The entire surface of the bottom glass plate is transparent and conductive. The bottom glass plate is called the backplane. A voltage between a segment contact (top plate) and the backplane contact (bottom plate) darkens the segment. To understand the operation in more detail, begin with the polarizer. Polarizing filters are attached to each glass plate. Polarizing filters can be arranged to either pass or block light. Figure 12-2 explains how the filters work. To complete the picture, place the glass plates and the nematic fluid in between the perpendicular polarizing filters. Refer to Figure 12-3. The light passes through the top polarizer, the molecules of the nematic fluid twist the light, and the light passes through the bottom polarizer. Thus, an LCD segment will look clear (or white) when no voltage is applied between the segment contact and the backplane contact.

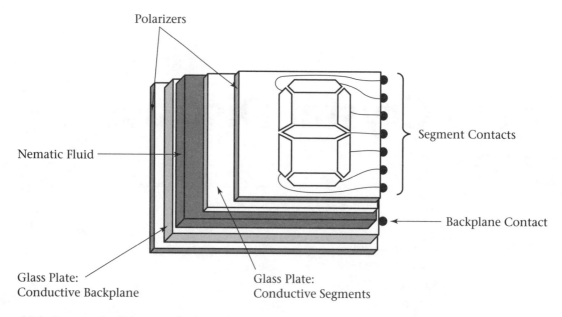

Figure 12-1 Construction Diagram of a Seven-Segment LCD

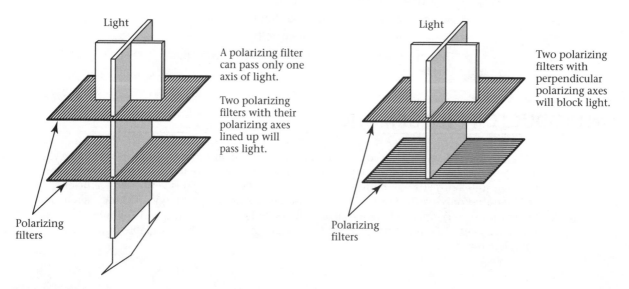

Figure 12-2 LCD Polarizing Filter

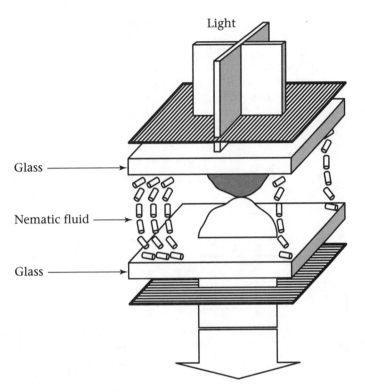

Figure 12-3 LCD without voltage passes light.

Figure 12-4 shows how a voltage applied between a segment contact and the backplane causes the nematic fluid to align its molecules. The light no longer twists and is blocked by the bottom polarizer. Thus, an LCD segment will look black (or dark) when voltage is applied between the segment contact and the backplane contact.

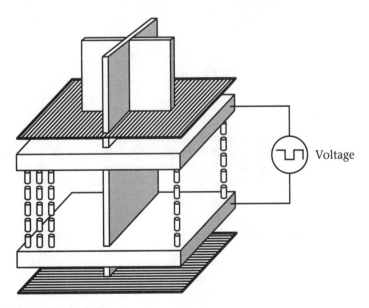

Figure 12-4 LCD with voltage blocks light.

The reason a pulse waveform must be used with an LCD is that a direct voltage (DC) will destroy the nematic fluid over time and shorten the life of an LCD. The optimum PPS rate of the pulse waveform is 30. There are two reasons why 30 PPS is optimum. The first reason is energy efficiency. At 30 PPS the energy used by the LCD is minimal. An LCD is like a capacitor. The conductive pattern on the upper glass plate is separated by the nematic fluid (capacitor dielectric) from the conductive backplane on the lower glass plate. Capacitors are short circuits to high-frequency signals and open circuits to low-frequency signals. Open circuits block current flow and, thus, reduce the energy consumption. The second reason why 30 PPS is optimal has to do with display flicker. The LCD display will flicker if the PPS rate is lower than 30. The human eye cannot detect flicker when the PPS rate is greater than 30. Thus, 30 PPS offers the most energy-efficient display that will *not* flicker.

Seven-Segment LCD Decoder System

Figure 12-5 shows how to connect a seven-segment LCD to a display decoder using XOR gates.

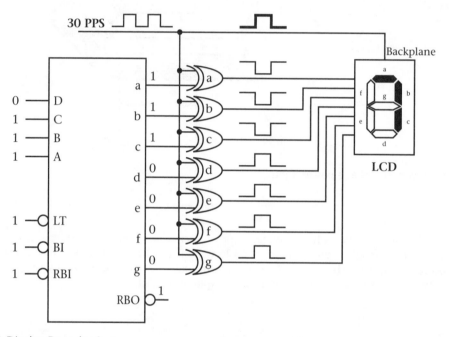

Figure 12-5 LCD Display Decoder System

In short, the XOR gates ensure that a pulse waveform is driving the LCD. XOR gates are controlled inverters. When a "1" is output from the display decoder, then the XOR passes the **inverse** of the 30 PPS pulse waveform. This **out-of-phase** pulse waveform creates a voltage across the top and bottom glass plates and darkens the liquid. When a "0" is output from the display decoder, then the XOR passes the "30 PPS" pulse waveform noninverted. This **in-phase** pulse waveform does not create a voltage across the top and bottom glass plates and the liquid remains clear.

Display Technology Evolution

Display technology evolved from numeric displays to alphanumeric displays. Alphanumeric displays can display both letters and numbers. For numeric-only information the seven-segment display works well. For alphanumeric information, a display with more segments is required. The next link in the evolutionary chain were 9-segment and then 16-segment displays. Refer to Figure 12-6.

A problem with 9- and 16-segment displays is the inability to display both capital and small letters. A dot matrix display (DMD) solves this problem. DMDs use round-shaped segments (also called **pixels**) that are arranged in a row and column pattern. Refer to Figure 12-7.

9-Segment Displays

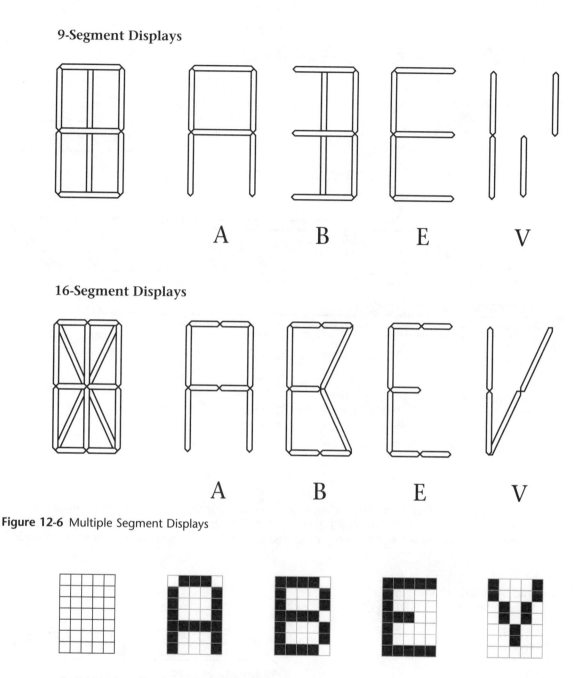

16-Segment Displays

A   B   E   V

Figure 12-6 Multiple Segment Displays

5x7 Matrix—5 columns and 7 rows of dots

Figure 12-7 5x7 Dot Matrix Display

Dot Matrix Displays (DMD) Using the HD44780 Controller: DMD Fundamentals

Many companies manufacture a dot matrix display (DMD) using the **Hitachi HD44780** controller IC. The HD44780 can control multiple DMDs using a structured command language. A structured command language simplifies the task of controlling a DMD. For example, there is a command that can be used to *clear* the entire display. There is a command that can be used to **shift** a message across the display. Figure 12-8 shows a two-line 16-character DMD.

Figure 12-8 shows that the DMD has 14 contacts or pins. Eight of the 14 pins are used for an **8-bit data bus**. The **data bus** is used to transfer two types of information: **DMD commands** (or

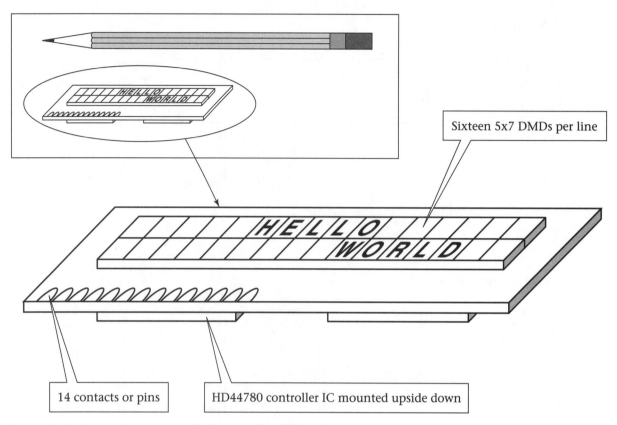

Figure 12-8 Two-line 16-character DMD using the Hitachi HD44780 controller

instructions) and **DMD data**. DMD commands include clear display, cursor on/off, display on/off, character position, and shift display. DMD data is a code that represents a letter of the alphabet. Figure 12-9 shows how a ROM can be connected to a DMD to display the message "Hello world."

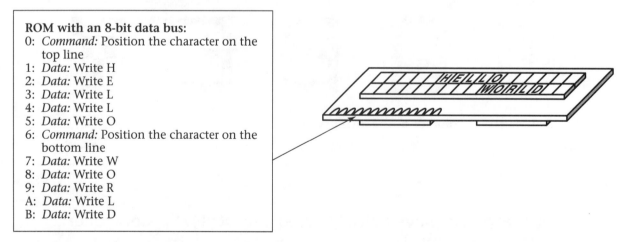

Figure 12-9 DMD displays the message "Hello world."

The ROM sends an 8-bit code representing a DMD command or DMD data. The DMD responds to the DMD command or DMD data to display the message. If the transfer rate from the ROM to the DMD is at a slow PPS rate, then you would see each letter of the message being spelled out. So how does the DMD know if the 8-bit number is a DMD command or DMD data?

Studying the complete function of all 14 DMD pins will give you the answer. Figure 12-10 shows the DMD pin out.

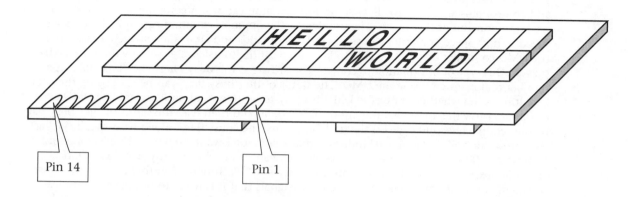

Pin #	Symbol	Function
1	Vss	*Power Supply:* Ground (0)
2	Vcc	*Power Supply:* 5 volts (1)
3	Vee	*Contrast Adjustment Input:* 0 volts = dark characters. 5 volts = light characters.
4	RS	*Register Select Input:* 1 = display data. 0 = command.
5	R/W	*Read/Write Input:* 1 = read the Busy Flag and 0 = write command or display data.
6	E	*Enable Input:* Negative edge triggered clock input.
7	DB0	LSB
8	DB1	
9	DB2	
10	DB3	*8-Bit Data Bus:* DB0 to DB7
11	DB4	
12	DB5	
13	DB6	
14	DB7	MSB

Note: Although the data bus can be used in 4-bit or 8-bit mode, this book will always use an 8-bit data bus.

Figure 12-10 DMD Pin Description

Figure 12-10 shows that **RS** is pin number **4**. *RS* is the **register select** input. The DMD records the 8-bit number from the data bus into either the command register or into the data register. **RS = 0** indicates that the 8-bit number from the data bus is a command. **RS = 1** indicates that the 8-bit number on the data bus is **display data**. "RS" is like a traffic controller directing the transfer of commands and display data.

Figure 12-10 shows that **E** is pin number **6**. *E* is the **enable** input. It is a negative edge triggered clock input. To complete a transfer, you must apply an 8-bit number onto the data bus, you must apply a "1" or "0" to "RS" and "R/W", and, finally, you must apply a negative edge to "E".

Figure 12-10 shows that **R/W** is pin number **5**. *R/W* is the **read/write** control input. **R/W = 0** indicates the 8-bit number on the data bus is being *written* into the DMD. **R/W = 1** indicates an 8-bit number is being *read* from the DMD data bus. To understand what 8-bit number can be read from the data bus, you will need to understand how a DMD communicates with a computer.

DMD and Computer Communications

The DMD uses an 8-bit data bus to receive commands and display data. The DMD data bus pins and control pins are set up to allow computer communications. A computer can be used to send a DMD command or DMD data. The problem with computer communication is the transfer speed. A high-speed computer can send DMD commands and DMD data at such a fast rate that the DMD controller may not be able to keep up with the transfer rate. So how does the DMD prevent this type of information overload? The answer is the **Busy Flag.** The "busy flag" is a number that is stored inside the DMD. The status of the "busy flag" can be read using the data bus. It indicates whether or not the DMD is ready to resume communications. The DMD expects a computer to write a DMD command (or DMD data) and then read the status of the "busy flag" before sending a second code on the data bus. Thus, the "busy flag," along with the "R/W," act like traffic controllers. "R/W = 0" indicates that a computer wants to write a DMD command or DMD data. "R/W = 1" indicates that a computer wants to read the "busy flag." The DMD controller expects the computer to **poll** the "busy flag." **Polling** implies that a computer must repeatedly read the "busy flag" until the DMD signals that it is ready to resume communications. Violating the status indicated by the "busy flag" results in an unstable display with unpredictable effects on the messages.

Displaying Data on a DMD

How does the DMD display data? It is a three-step process. First, an 8-bit "ASCII" data code is written to the DMD data bus. Second, the DMD uses the **ASCII** code to look up the pixel data for the character in an internal ROM called **Character Generator ROM (CGROM).** Third, the DMD transfers the pixel data from CGROM to the **Display Data RAM (DDRAM).** The content of DDRAM gets displayed on the DMD. Refer to Figure 12-11.

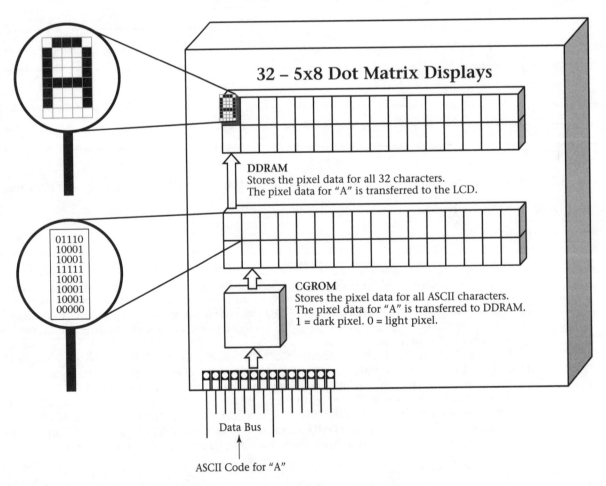

32 – 5x8 Dot Matrix Displays

DDRAM
Stores the pixel data for all 32 characters.
The pixel data for "A" is transferred to the LCD.

CGROM
Stores the pixel data for all ASCII characters.
The pixel data for "A" is transferred to DDRAM.
1 = dark pixel. 0 = light pixel.

Data Bus

ASCII Code for "A"

Figure 12-11 DMD displays the letter *A*

ASCII is an abbreviation for the words **American Standard Code for Information Interchange**. It is pronounced **AS-KEY**. It is a 7-bit code that is used to represent alphanumeric information. A 7-bit code has 2^7 or 128 different binary combinations or codes.

- 26 binary codes are used to represent small letters a through z.
- 26 binary codes are used to represent capital letters A through Z.
- 10 binary codes are used to represent the numbers 0 through 9.
- The remaining binary codes are used to represent punctuation and special characters (:, ;, $, ! ? ...).

ASCII is an alphanumeric code that is widely accepted as an industry standard.

"**Character Generator ROM**" is a nonvolatile memory system that stores the pixel data for all the ASCII characters. **Nonvolatile** means that the DMD is always ready to display the ASCII characters even after a power shutdown. When an ASCII code for DMD data is written to the data bus, the pixel data for that character is retrieved from the CGROM and transferred to the DDRAM.

"**Display Data RAM**" is a volatile memory system that stores the pixel data for each character that is being displayed. RAM allows the display information to be altered whenever a message needs to be changed. The contents of RAM are lost when the power is shut down. When an ASCII code for DMD data is written to the data bus, the pixel data for that character is retrieved from the CGROM and transferred to the DDRAM. DDRAM stores the pixel data for all the characters of the current message being displayed.

DMD CGROM

Figure 12-12 shows the pixel data stored inside the DMD "CGROM."

CGROM stores the pixel data for the American and the Japanese alphabet. For example, according to the CGROM chart, capital *A* is a lower 4-bit number = **xxxx0001** and a higher 4-bit number = **0100**. The higher 4-bit number "0100" replaces the "xxxx" in the lower 4-bit number creating an 8-bit number 0100 0001 (or **41** HEX). The 8-bit number is the 7-bit ASCII code for the letter with the **MSB = 0**. If you study the table in Figure 12-12 you will notice that all the American letters of the alphabet are accessed in CGROM with **0** as the **MSB**. ASCII is a 7-bit code and does not use the eighth bit. Hitachi uses the eighth bit to include the letters of their native country. The characters for Japanese alphabet all have the eighth (MSB) bit set to 1. The open column under **higher 4-bit = 0000** is blank. This area is set aside for CGRAM. **CGRAM** allows the creation of **custom characters**. CGRAM and custom characters are explained later.

Figure 12-12 DMD CGROM

Example: Capital A:

Higher 4-bit = 0100
Lower 4-bit = xxxx 0001

8-bit ASCII = 0100 0001

The higher 4-bit number replaces the xxxx of the lower 4-bit number to create an ASCII code.

DMD DDRAM

DDRAM is used to store the pixel data for the current message being displayed. As display data is being written into the DMD, the pixel data that makes up the message is being transferred from CGROM into the DDRAM. How does the DMD allow you to select a starting position for a message? What if you would like to display two words with one word centered on the top line and the other word centered on the bottom line? The DMD uses an address code to identify the position of each character in the DDRAM. Refer to Figure 12-13.

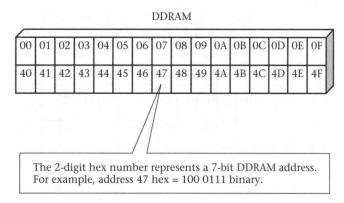

Figure 12-13 DMD DDRAM Addressing

The DMD has a command that allows you to select any address as the starting position for your message. Figure 12-14 shows you how to display a two-word message with one word centered on the top line and the other word centered on the bottom line.

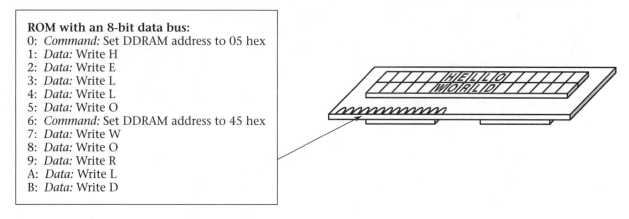

Figure 12-14 Setting the DMD DDRAM Address

DDRAM can store 24 **offscreen characters** per line. Offscreen characters can be used to store a message and later scroll that message into view. It allows the DMD to act like a **scrolling billboard**. A message can be displayed in the viewable area of DDRAM while a second message is written offscreen. The DMD has a **shift display** command that can be used to move the first message offscreen while the second message is scrolled into view. Refer to Figure 12-15.

Visually, DDRAM is laid out like a **cylinder**. The end of the offscreen characters are visually positioned next to the first characters. Figure 12-16 shows how a message written at the beginning of the viewable area is next to the message written at the end of the offscreen area.

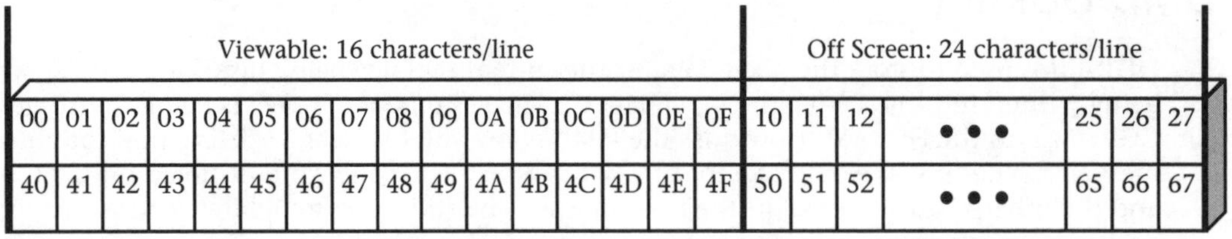

Viewable: 16 characters/line
Off Screen: 24 characters/line
00 01 02 03 04 05 06 07 08 09 0A 0B 0C 0D 0E 0F 10 11 12 • • • 25 26 27
40 41 42 43 44 45 46 47 48 49 4A 4B 4C 4D 4E 4F 50 51 52 • • • 65 66 67

Figure 12-15 DMD DDRAM Offscreen Characters

Viewable: 16 characters/line
Off Screen: 24 characters/line
00 01 D 02 A 03 Y 04 05 06 07 08 09 0A 0B 0C 0D 0E 0F 10 11 12 • • • 24 G 25 O 26 O 27 D
40 41 42 43 44 45 46 47 48 49 4A 4B 4C 4D 4E 4F 50 51 52 • • • 64 65 66 67
24 G 25 O 26 O 27 D 00 01 D 02 A 03 Y
64 65 66 67 40 41 42 43
DDRAM

Figure 12-16 DMD DDRAM Cylinder Effect

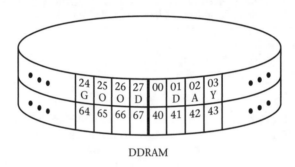

DMD CGRAM

Character Generator RAM can be used to create eight custom characters. Custom characters can be used to create symbols for letters of the alphabet that are not part of the American and Japanese languages. Custom characters can also be used to create graphic symbols. Figure 12-17 shows the pixel data for a "happy face" symbol. Each custom character uses an identification label. The first custom character is **CG0: Character Generator 0**. The eighth custom character is **CG7**. To create a custom character, you must transfer the pixel data to CGRAM. Figure 12-18 shows a "Happy Face" character and "Sad Face" character stored in CGRAM.

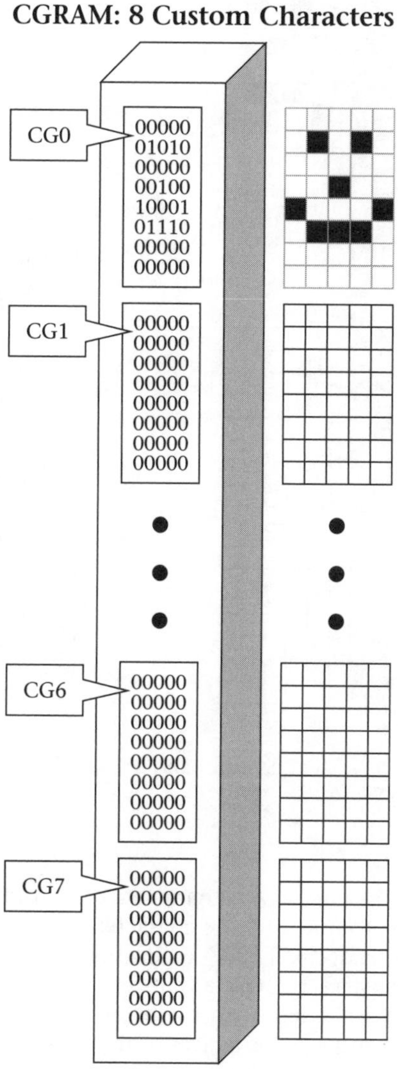

CGRAM: 8 Custom Characters

Figure 12-17 DMD CGRAM "Happy Face" Custom Character

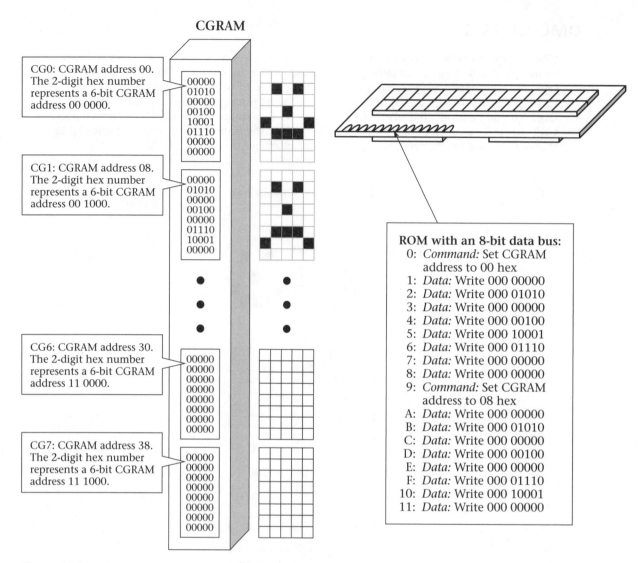

Figure 12-18 DMD CGRAM with a "Happy Face" Symbol and a "Sad Face" Symbol

CG0 is used to store the happy face character. The command **CGRAM address set** is used to initialize the pixel data that represents the character. The starting address for "CG0" is a 6-bit number that represents hexadecimal "00." Following the "CGRAM address set" command are eight consecutive data transfers. Each data transfer identifies a row of pixels. Each data transfer is an 8-bit number in which 5 bits represent the pixel data and the 3 most significant bits are not used. A "1" is a dark pixel and a "0" is a clear or light pixel.

CG1 is used to store the sad face character. Because eight rows of pixel data must be written to initialize a character, each CGRAM starting address is separated by eight address positions. Thus, the starting address for "CG1" is a 6-bit number that represents hexadecimal "08." Following the "CGRAM address set" command are, once again, eight consecutive data transfers. Once again, each data transfer identifies a row of pixels.

Each row of pixels in a CGRAM character can be separately changed. For example, if you wanted to separate the distance between the eyes of the sad face character, you could use the "CGRAM address set" command with a 6-bit number that represents hexadecimal "09" and then follow that with a data transfer **Data: Write 000 10001**. This allows changes to a character to be made without having to retransfer all eight rows of pixel data.

To display (or access) a custom character requires a different mode of character addressing. You must use the address represented by the 8-bit numbers shown in Figure 12-19.

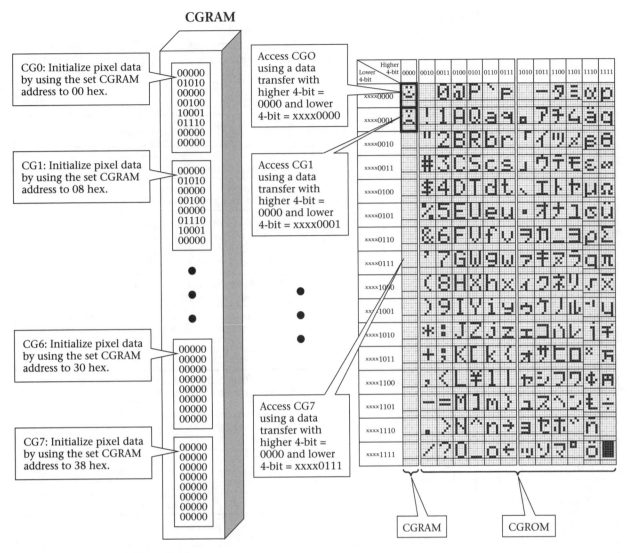

Figure 12-19 DMD CGRAM Table

Custom characters present a challenge because they use two levels of addressing. As you have seen, one level is used to store the pixel data for the custom character. The second level is used to access the custom character. The custom characters shown in Figure 12-19 occupy a column in the table that is adjacent to the CGROM characters. The address used to display a custom character "CG0" (happy face) is generated by performing a data transfer with "Higher 4-bit" = "0000" and "lower 4-bit" = "xxxx 0000." A data transfer using "Higher 4-bit" = "0000" and "lower 4-bit" = "xxxx 0001" will display the sad face character. A sample program is presented later in this lab that will demonstrate all CGRAM concepts.

DMD Commands Summary

The DMD commands are shown in Figure 12-20. Most commands have different control effects based on their bit options. Later in the lab, sample programs are presented to demonstrate their effect.

COMMAND	R/W	RS	DB7	DB6	DB5	DB4	DB3	DB2	DB1	DB0	DESCRIPTION
CLEAR DISPLAY	0	0	0	0	0	0	0	0	0	1	Clears display and returns cursor to home position (DDRAM address = 00H).
SEND CURSOR HOME	0	0	0	0	0	0	0	0	1	*	Returns cursor to home position. Returns a shifted display to original position. Data in DDRAM is not changed.
SET THE DATA WRITE MODE	0	0	0	0	0	0	0	1	I/D	S	I/D sets the CURSOR/DISPLAY move direction. The CURSOR/DISPLAY will automatically move to the next DDRAM/CGRAM address after each data write. S specifies to shift the display or to move the cursor.
SET DISPLAY/ CURSOR ON/OFF	0	0	0	0	0	0	1	D	C	B	D sets display On/Off. C sets the cursor On/Off. B enables the character at the cursor position to Blink/Not Blink.
SHIFT/MOVE CURSOR/DISPLAY	0	0	0	0	0	1	S/C	R/L	*	*	S/C moves the cursor or shifts the display. Contents of DDRAM are not changed. R/L chooses the direction of the Shift/Move.
SET DISPLAY MODE (DISPLAY FUNCTION)	0	0	0	0	1	DL	N	F	*	*	DL sets the Data Bus size. N sets the number of lines to be used. F sets the Dot Matrix font (first instruction).
SET CGRAM ADDRESS	0	0	0	1		ACG (6-bit address)					Sets the starting CGRAM address. The CGRAM data is sent to this starting address after this setting.
SET DDRAM ADDRESS	0	0	1			ADD (7-bit address)					Sets the starting DDRAM address. The DDRAM ASCII/Japan/Custom characters are sent to this starting address after this setting.
READ BUSY FLAG/ADDRESS	1	0	BF			AC					Read the Busy Flag (BF) indicating internal operation is being performed and reads address counter contents.
CGRAM/DDRAM DATA WRITE	0	1			8-bit code for ASCII/Japan/Custom Characters						Writes data into DDRAM or CGRAM. Used after address has been set. The address pointer is inc/dec automatically.
READ DDRAM/ CGRAM DATA	1	1			Read Data						Read data from DDRAM or CGRAM.

I/D = 1: Increment the DDRAM address.
  0: Decrement the DDRAM address.

S = 1: Shift the display after data write. Cursor freeze.
  0: Move the cursor after data write. Display freeze.

D = 1: Turn the entire display ON.
  0: Turn the entire display OFF.

C = 1: Turn the cursor ON.
  0: Turn the cursor OFF.

B = 1: Blink the character at the cursor position.
  0: DO NOT blink the character at the cursor position.

S/C = 1: Shift the display.
  0: Move the cursor.

R/L = 1: Shift/Move to the RIGHT.
  0: Shift/Move to the LEFT.

DL = 1: Use an 8-bit data bus.
  0: Use a 4-bit data bus.

N = 1: Use 2 lines of the display.
  0: Use 1 line of the display.

F = 1: 5x10 dot matrix font.
  0: 5x7 dot matrix font.

BF = 1: Busy doing an internal operation.
  0: Ready to communicate.

DDRAM: Display Data RAM.
CGRAM: Character Generator RAM.

ACG: CGRAM Address
ADD: DDRAM Address

AC: CGRAM/DDRAM Address Counter
*: Not used. Set it to 0.

Figure 12-20 DMD Commands Summary

A ROM-Based Electronic Billboard System

A ROM-Based Electronic Billboard System: Construction Diagram

At this point you have seen how a DMD uses CGROM, DDRAM, and CGRAM to display messages. It is time to create a basic messaging system with the DMD. An electronic billboard system is capable of displaying scrolling messages. Although the HD44780 DMD is intended to be computer controlled, a simple ROM-based system can be used to control it. A construction diagram for an **electronic billboard system** is shown in Figure 12-21.

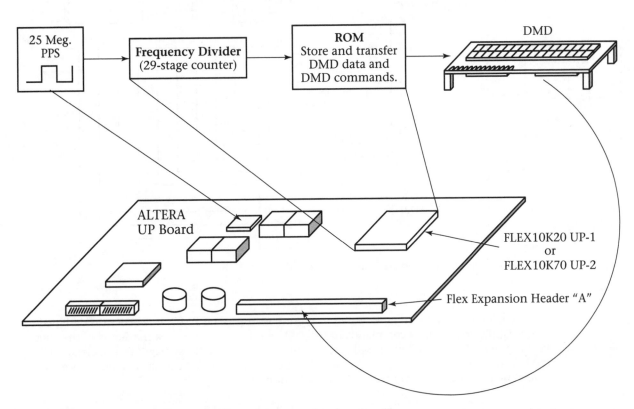

Figure 12-21 ROM-Based Electronic Billboard System: Construction Diagram

The basic operation of the messaging system uses a frequency divider (29-stage counter) to slow down the 25 Meg PPS clock and to cycle the address inputs of the ROM. The ROM stores the codes for DMD data and DMD commands. The ROM writes the codes to the DMD to display the messages. The DMD is connected to FLEX expansion header. Appendix C describes the FLEX expansion header in more detail.

A ROM-Based Electronic Billboard System: System Diagram

Figure 12-22 shows a system diagram for the electronic billboard system. The DMD diagram now includes blocks that represent CGRAM and the command register. The CGRAM block is used to store pixel data for custom characters. Refer to the section "DMD CGRAM" presented earlier in the lab for CGRAM operational details. The command register is used to store and execute DMD commands like clear display, shift display, and so on, when "RS = 1." DMD data is sent to CGROM and CGRAM area when "RS = 0."

The **contrast** adjust input can be hooked up to a potentiometer (variable resistor) to allow contrast to be adjusted; however, it simplifies the design if it is directly connected to ground (0 volt). Ground makes the characters as dark as possible and offers the best character visibility in a bright room.

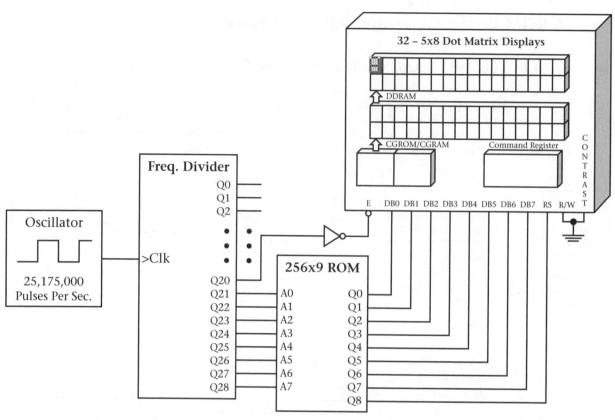

Figure 12-22 ROM-Based Electronic Billboard System: System Diagram

The R/W input is directly connected to ground (0 volt). The DMD, however, was designed to receive DMD data or DMD commands in the write mode and then check the status of the "busy flag" in the *read* mode. A ROM-based system can never read the "busy flag." To understand how a ROM-based system can be made to work without being able to read the "busy flag," you must review the information presented in the next section.

A ROM-Based Electronic Billboard System: System Communication

In the previous section titled "DMD and Computer Communications" you saw how the HD44780 DMD is designed to allow computer communications. The DMD expects a computer to *write* a DMD command (or DMD data) and then *read* (or poll) the status of the "busy flag" before sending a new 8-bit code on the data bus. Violating the status indicated by the "busy flag" results in an unstable display with unpredictable effects on messaging. The obstacle that a ROM based system must overcome is its inability to read the "busy flag." The solution to the problem is evident if you compare the transfer rate of the ROM-based system to the transfer rate specifications of the DMD.

Q20 through **Q28** of the frequency divider (29-stage counter) are used to address the ROM and clock the DMD. Figure 12-23 shows the timing diagram of the frequency divider. The NOT gate inverts the "Q20" waveform and ensures that the DMD is clocked when the ROM data is stable (not changing). When "Q20" changes from "1" to "0," it cycles the count sequence generated at outputs "Q21" to "Q28." A changing count sequence, in turn, cycles the address inputs of the ROM. When the ROM addresses are changing, the data out of the ROM is not stable. Although the amount of time required for the ROM data to become valid is very insignificant, the NOT gate delays the "E" input from transferring the data by half a count cycle. This delay ensures that the ROM data is valid.

The "Q20" clock rate is "12 PPS." The address inputs of the ROM change every 83 milliseconds (1/12). DMD data or DMD commands are transferred to the DMD every 83 milliseconds. This 83-millisecond transfer rate will become an important part of the solution.

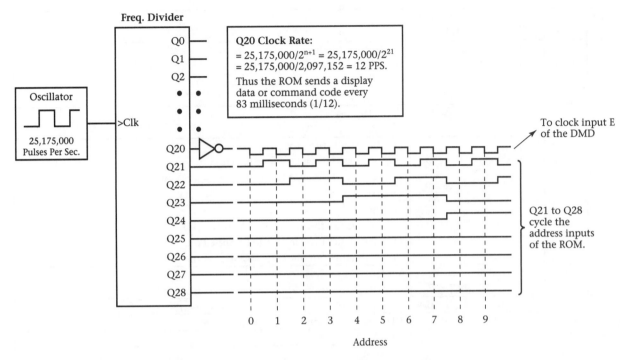

Figure 12-23 ROM-Based Electronic Billboard System: Counter Timing Diagram

Most DMD data and DMD command write operations have an execution time of 40 microseconds. This means it takes the DMD about 40 microseconds to execute the transfer. During this time interval, the DMD is busy and cannot receive any new DMD data or DMD commands. The "Clear Display" and "Home Cursor" commands have the longest execution time of all the write operations. They each have an execution time of 1.64 milliseconds. A computer connected to the DMD can poll the "busy flag" before sending a new DMD data or DMD command. A ROM-based display system can use the fact that the longest execution time is 1.64 milliseconds to avoid reading the "busy flag." The transfer rate from the ROM to the DMD is 83 milliseconds. This exceeds the longest execution time of 1.64 milliseconds by about 81 milliseconds. This 81-millisecond time interval guarantees the DMD is ready without any need to read the "busy flag." As long as the transfer rate from the ROM is slow enough to guarantee the execution of any and all transfers, there is no need to read the "busy flag."

Sample Program #1: Display a Two-Word Message

The ROM-based system will be used to display a two-word message "Hello World." The first word "Hello" will be displayed in the middle of the first line in the viewable area of the DMD. The second word "World" will be written offscreen on the second line and scrolled into view. The word "World" will appear in the middle of the second line. Figure 12-24 shows a table with the sample program.

HEX Address: 0 to 4

These commands do not display any characters. They are used to *initialize* the DMD. These commands get the DMD ready to display data, from left to right, starting at DDRAM address "5 HEX" without a visible cursor.

HEX Address: 5 to 9

Display the message "Hello" in the middle of the first line without a visible cursor. You will see each letter of the word "Hello" appear one character at a time.

HEX Address: A to F

The message "World" is written offscreen at the end of the visible area of the bottom line. You will not see this message being written.

HEX Address: 10 to 19

Hex Add	Hex Data	RS Q8	DB7... Q7...Q4	...DB0 Q3...Q0	Description
0	0038	0	0011	1000	**Set Display Mode:** - D/L = 1 specifies an 8-bit data bus. - N = 1 specifies to use 2 lines on the DMD. - F = 0 for a 5x7 DMD.
1	0001	0	0000	0001	**Clear Display**
2	0006	0	0000	0110	**Set the Data Write Mode:** - I/D = 1 sets the display move direction to automatically increment to the next DDRAM/CGRAM address after each data write. - S = 0 specifies to move the cursor after a data write operation.
3	000C	0	0000	1100	**Set Diplay/Cursor ON/OFF** - D = 1 specifies the display is to be turned on. - C = 0 specifies the cursor is to be turned off. - B = 0 specifies not blink the character at the cursor position.
4	0085	0	1 000	0101	**Set DDRAM Address:** - 000 0101 specifies address 05 hex. Middle of top line.
5	0148	1	0100	1000	**DDRAM Data Write:** - 0100 1000 specifies "H"
6	0165	1	0110	0101	**DDRAM Data Write:** - specifies "e"
7	016C	1	0110	1100	**DDRAM Data Write:** - specifies "l"
8	016C	1	0110	1100	**DDRAM Data Write:** - specifies "l"
9	016F	1	0110	1111	**DDRAM Data Write:** - specifies "o"
A	00D0	0	1 101	0000	**Set DDRAM Address:** - 101 0000 specifies address 50 hex. First off-screen character of bottom line.
B	0157	1	0101	0111	**DDRAM Data Write:** - 0101 0111 specifies "W"
C	016F	1	0110	1111	**DDRAM Data Write:** - specifies "o"
D	0172	1	0111	0010	**DDRAM Data Write:** - specifies "r"
E	016C	1	0110	1100	**DDRAM Data Write:** - specifies "l"
F	0164	1	0110	0100	**DDRAM Data Write:** - specifies "d"
10	0018	0	0001	1000	**Shift Display:** - S/C = 1 specifies shift display - R/L - 0 specifies left.
11	0018	0	0001	1000	**Shift Display:** - S/C = 1 specifies shift display - R/L - 0 specifies left.
12	0018	0	0001	1000	**Shift Display:** - S/C = 1 specifies shift display - R/L - 0 specifies left.
13	0018	0	0001	1000	**Shift Display:** - S/C = 1 specifies shift display - R/L - 0 specifies left.
14	0018	0	0001	1000	**Shift Display:** - S/C = 1 specifies shift display - R/L - 0 specifies left.
15	0018	0	0001	1000	**Shift Display:** - S/C = 1 specifies shift display - R/L - 0 specifies left.
16	0018	0	0001	1000	**Shift Display:** - S/C = 1 specifies shift display - R/L - 0 specifies left.
17	0018	0	0001	1000	**Shift Display:** - S/C = 1 specifies shift display - R/L - 0 specifies left.
18	0018	0	0001	1000	**Shift Display:** - S/C = 1 specifies shift display - R/L - 0 specifies left.
19	0018	0	0001	1000	**Shift Display:** - S/C = 1 specifies shift display - R/L - 0 specifies left.

Figure 12-24 ROM-Based Electronic Billboard System: Sample Program #1

Ten **Shift Display Left** commands scroll the message "World" into the middle of the second line while the message "Hello," on the first line, scrolls offscreen. Figure 12-25 shows the DMD before and after the 10 **Shift Display Left** commands.

The **Shift Display Left** command moves the viewable area one character position at a time. The DDRAM display data does not move within the DDRAM memory; it is the viewable window that moves.

Sample Program #2: Display a "Happy Face" Custom Character

The electronic billboard system will be used to display a "Happy Face" custom character. The pixel data will be placed in "CG1." Figure 12-26 shows a table with the sample program.

Viewable: 16 characters/line	Off Screen: 24 characters/line

| 00 | 01 | 02 | 03 | 04 | 05 H | 06 e | 07 l | 08 l | 09 o | 0A | 0B | 0C | 0D | 0E | 0F | 10 | 11 | 12 | 13 | 14 | ••• | 25 | 26 | 27 |
| 40 | 41 | 42 | 43 | 44 | 45 | 46 | 47 | 48 | 49 | 4A | 4B | 4C | 4D | 4E | 4F | 50 W | 51 o | 52 r | 53 l | 54 d | ••• | 65 | 66 | 67 |

DMD before the shift display left commands

Off Screen:	Viewable: 16 characters/line	Off Screen:

| 05 H | 06 e | 07 l | 08 l | 09 o | 0A | 0B | 0C | 0D | 0E | 0F | 10 | 11 | 12 | 13 | 14 | 15 | 16 | 17 | 18 | 19 | ••• | 25 | 26 | 27 |
| 45 | 46 | 47 | 48 | 49 | 4A | 4B | 4C | 4D | 4E | 4F | 50 W | 51 o | 52 r | 53 l | 54 d | 55 | 56 | 57 | 58 | 59 | ••• | 65 | 66 | 67 |

DMD after the shift display left commands

Figure 12-25 ROM-Based Electronic Billboard System: Sample Program #1 messaging results

Hex Add	Hex Data	RS Q8	DB7... Q7...Q4	...DB0 Q3...Q0	Description
0	0038	0	0011	1000	**Set Display Mode:** - D/L = 1 specifies an 8-bit data bus. - N = 1 specifies to use 2 lines on the DMD. - F = 0 for a 5x7 DMD.
1	0001	0	0000	0001	**Clear Display**
2	0006	0	0000	0110	**Set the Data Write Mode:** - I/D = 1 sets the display move direction to automatically increment to the next DDRAM/CGRAM address after each data write. - S = 0 specifies to move the cursor after a data write operation.
3	000C	0	0000	1100	**Set Diplay/Cursor ON/OFF** - D = 1 specifies the display is to be turned on. - C = 0 specifies the cursor is to be turned off. - B = 0 specifies not blink the character at the cursor position.
4	0048	0	01 00	1000	**Set CGRAM Address:** - 00 1000 specifies address 08 hex for CG1.
5	0100	1	000 0	0000	Top row of pixels ⭕⭕⭕⭕⭕
6	010A	1	000 0	1010	Next row of pixels ⭕⬤⭕⬤⭕
7	0100	1	000 0	0000	Next row of pixels ⭕⭕⭕⭕⭕
8	0104	1	000 0	0100	Next row of pixels ⭕⭕⬤⭕⭕
9	0111	1	000 1	0001	Next row of pixels ⬤⭕⭕⭕⬤
A	010E	1	000 0	1110	Next row of pixels ⭕⬤⬤⬤⭕
B	0100	1	000 0	0000	Next row of pixels ⭕⭕⭕⭕⭕
C	0100	1	000 0	0000	Next row of pixels ⭕⭕⭕⭕⭕ Cursor ROW (descender)
D	0085	0	1 000	0101	Set DDRAM Address: - 000 0101 specifies address 05 hex. Middle of top line.
E	0101	1	0000	0001	Display the custom character.

Figure 12-26 ROM-Based Electronic Billboard System: Sample Program #2: CGRAM Programming

The **Set CGRAM Address** command uses a 6-bit number for the **pixel data** address. Subsequent data transfers are sent as pixel data to CGRAM to initialize the custom character. The

Set DDRAM Address command reverts the DMD to transfer data and stops the custom character pixel data transfers.

> *NOTE: If you do not use the command "Set DDRAM Address," after CGRAM pixel initialization, then subsequent data transfers will be stored in CGRAM as pixel data and will not appear as displayable CGROM characters stored in DDRAM.*

Figure 12-27 shows all eight CGRAM pixel initialization addresses and the eight character access addresses.

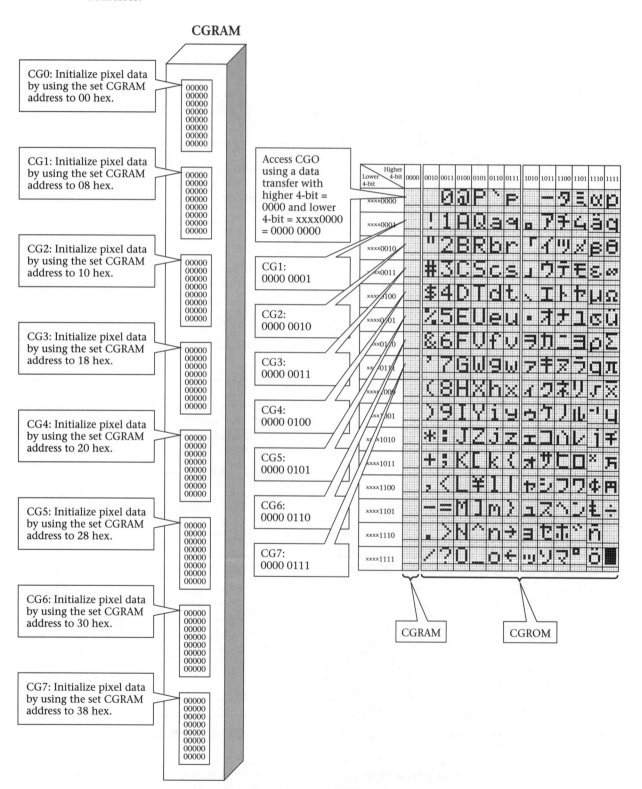

Figure 12-27 DMD CGRAM Addresses

POWERPOINT PRESENTATION

Use PowerPoint to view the Lab 12 slide presentation.

LAB WORK PROCEDURE

You will build and program an electronic billboard system using a DMD with a Hitachi controller and the Altera UP board.

Build an Electronic Billboard System

Although some ROM configurations work with the MAX IC, it is recommended that you use the FLEX IC. You will build a ROM-based electronic billboard system.

To complete this lab you need to connect a DMD with a Hitachi HD44780 controller to the FLEX expansion header. You will draw a system diagram that connects a 29-stage binary counter to the address inputs of a 256×9 ROM. The counter will sequence the address inputs of the ROM. The ROM data outputs will send **DMD data** and **DMD commands** to a DMD connected to the FLEX expansion header. You will test the operation of this electronic billboard system using the code for "Sample Program #1."

1$_{\text{FLEX}}$. You will draw a diagram for an electronic billboard system. Use the Lab Work Procedure in Lab 1 as a guide, and complete steps 1 through 12. Begin by creating a folder (directory) on your blank disk named **Labs**. Use the project name **Lab12**. The ROM symbol in the "mega_lpm" library is called **LPM_ROM**. The counter symbol in the "mega_lpm" library is called **LPM_COUNTER**. The ground symbol in the "prim" library is called **gnd**. The Graphic Editor file for the electronic billboard system should resemble Figure 12-28.

2$_{\text{FLEX}}$. The UP board can now be programmed. Lab 1 describes all the steps that are required to program a UP board with the electronic billboard system. Use the Lab Work Procedure in Lab 1 as a guide, and complete steps 13 through 14. New procedures for steps 15 through 17 are listed next.

Before compiling, the contents of ROM need to be initialized. The initialization comes from the **Memory Initialization File (MIF) a:\labs\lab12.mif**. To create this initialization file, you must *fill in a memory map diagram* and *enter the data into the MIF file*. The contents of the "memory map diagram" control the appearance of the message. When creating an original message, you would fill in your own "memory map diagram." For this lab procedure, you will enter the code for "Sample Program #1" shown in Figure 12-24. Upcoming lab exercises will have you create your own memory map diagram for your own original messages.

Enter the memory map data for "Sample program #1" into the MIF file:

There are two procedures that you can follow to enter the memory map diagram into the MIF file. **Method #1** uses the DMD editor. **Method #2** uses the Altera initialize memory window. Ask your instructor which method you should use.

Method #1: Use the DMDeditor provided on the CD-ROM.

The **DMDeditor** is a program provided on the CD-ROM that allows you to create, edit, and document a memory initialization file. It is like an MIF word processor. It greatly simplifies the process of creating code for the electronic billboard system.

a. The DMDeditor is set up to open the **lab12.mif** file located on the **current drive** and **current folder**. Thus, you must place the editor program in folder **Labs** of your floppy disk. Minimize the Altera MAX+PLUS II window and use "Windows Explorer" to copy the file **DMDeditor.exe** from the CD-ROM to folder **Labs** of your floppy disk.

b. You must create a blank **lab12.mif** file. This procedure need only be done once. Once the file is created you can skip this step and proceed to step c. There is no need to follow this procedure if you are trying to make changes to an existing **lab12.mif** file.

- Maximize the Altera MAX+PLUS II window.
- Complete the "Project Save & Compile" step. Ignore any warnings.
- From the **MAX+PLUS II** menu select **Simulator**. The simulator window will open. It is shown in **Lab 11**, Figure 11-27.
- From the menu that runs across the very top of the screen, click on **Initialize** then select **Initialize Memory**. The "Initialize Memory" window shown in Figure 12-33 will open.

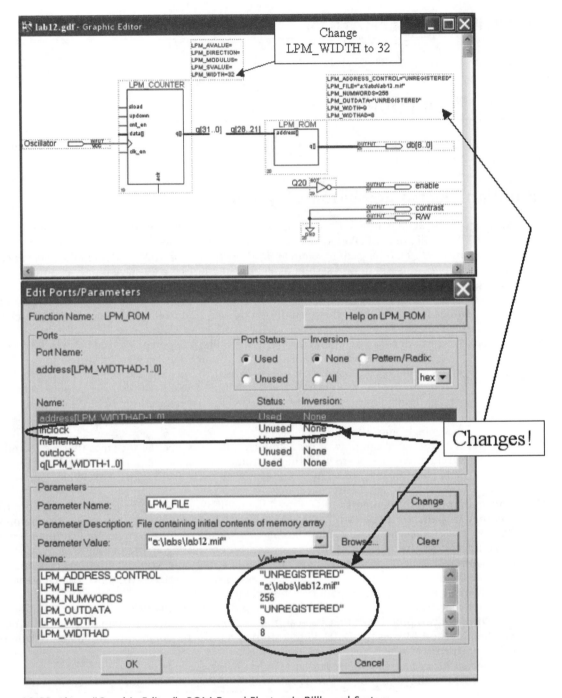

Figure 12-28 Altera "Graphic Editor": ROM-Based Electronic Billboard System

- Click the **Export File** button. Save the blank code (all data = 000) as file **a:\labs\lab12.mif**. Click "OK."

c. Use the DMDeditor to enter the code into the **lab12.mif** file. Minimize the Altera **MAX+Plus II** window and use "Windows Explorer" to double-click on the file **a:\labs\DMDeditor.exe**. This will run the DMDeditor program and you will see the window shown in Figure 12-29 open.

- From the **File** menu select **Open lab12.mif**. The DMDeditor window shown in Figure 12-30 will appear.
- Each line begins with a ROM HEX address and a ":" separating the HEX data "0000." If you scroll down the screen you will find the last address is **FF** (255). Use the DMDeditor to enter the memory map data for "Sample Program #1" shown in Figure 12-24. When you are done, the DMDeditor should resemble Figure 12-31.

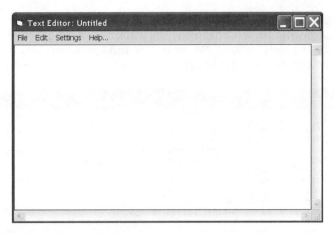

Figure 12-29 DMDeditor Window: Blank Window

```
Text Editor: lab12.mif
File  Edit  Settings  Help...
0       :         0000; --
1       :         0000; --
2       :         0000; --
3       :         0000; --
4       :         0000; --
5       :         0000; --
6       :         0000; --
7       :         0000; --
8       :         0000; --
9       :         0000; --
a       :         0000; --
b       :         0000; --
c       :         0000; --
d       :         0000; --
e       :         0000; --
f       :         0000; --
10      :         0000; --
```

Figure 12-30 DMDeditor Window: Lab12.mif

```
Text Editor: lab12.mif
File  Edit  Settings  Help...
0       :         0038; --
1       :         0001; --
2       :         0006; --
3       :         000C; --
4       :         0085; --
5       :         0148; --
6       :         0165; --
7       :         016c; --
8       :         016c; --
9       :         016f; --
a       :         00d0; --
b       :         0157; --
c       :         016f; --
d       :         0172; --
e       :         016c; --
f       :         0164; --
10      :         0018; --
11      :         0018; --
12      :         0018; --
13      :         0018; --
14      :         0018; --
15      :         0018; --
16      :         0018; --
17      :         0018; --
18      :         0018; --
19      :         0018; --
```

Figure 12-31 the DMDeditor Window: "Sample Program #1"

- Use the DMDeditor to document the code. The DMDeditor can comment each line of the **MIF** file automatically. The comments need never be typed in manually. From the File menu select √ **Fix addresses and comments** and the DMDeditor will add the comments. Refer to Figure 12-32.

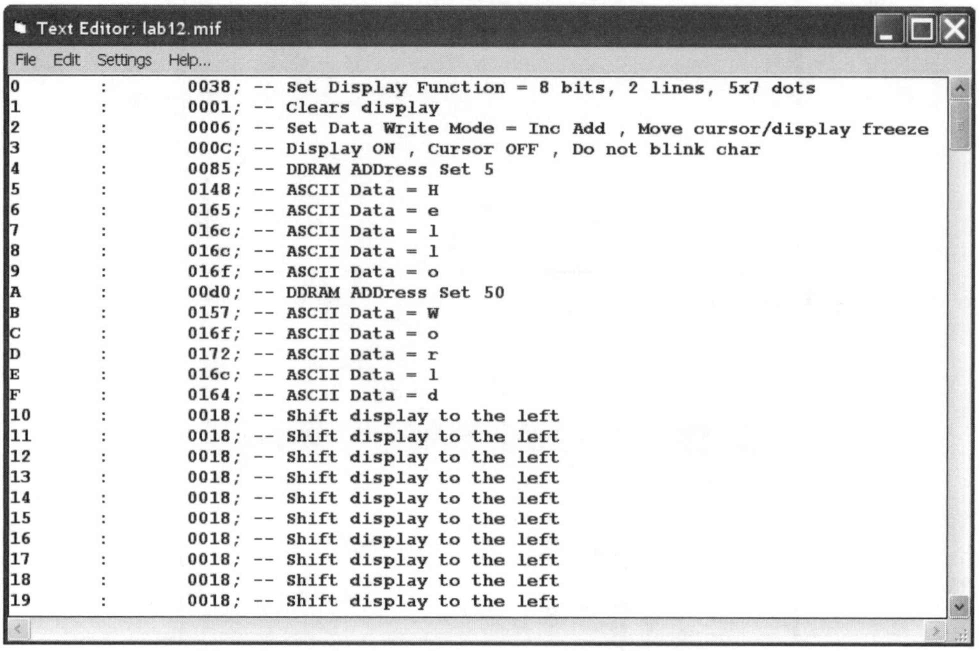

```
Text Editor: lab12.mif
File  Edit  Settings  Help...
0    :        0038; -- Set Display Function = 8 bits, 2 lines, 5x7 dots
1    :        0001; -- Clears display
2    :        0006; -- Set Data Write Mode = Inc Add , Move cursor/display freeze
3    :        000C; -- Display ON , Cursor OFF , Do not blink char
4    :        0085; -- DDRAM ADDress Set 5
5    :        0148; -- ASCII Data = H
6    :        0165; -- ASCII Data = e
7    :        016c; -- ASCII Data = l
8    :        016c; -- ASCII Data = l
9    :        016f; -- ASCII Data = o
A    :        00d0; -- DDRAM ADDress Set 50
B    :        0157; -- ASCII Data = W
C    :        016f; -- ASCII Data = o
D    :        0172; -- ASCII Data = r
E    :        016c; -- ASCII Data = l
F    :        0164; -- ASCII Data = d
10   :        0018; -- Shift display to the left
11   :        0018; -- Shift display to the left
12   :        0018; -- Shift display to the left
13   :        0018; -- Shift display to the left
14   :        0018; -- Shift display to the left
15   :        0018; -- Shift display to the left
16   :        0018; -- Shift display to the left
17   :        0018; -- Shift display to the left
18   :        0018; -- Shift display to the left
19   :        0018; -- Shift display to the left
```

Figure 12-32 DMDeditor Window: "Sample Program #1" with Comments

- Save **lab12.mif**, exit the DMDeditor, and return to the Altera MAX+PLUS II program. From the **File** menu select **Save "lab12.mif" and exit**. Maximize the Altera MAX+PLUS II window.

d. You need to compile the project again. The compiler takes the MIF information (**a:\labs\lab12.mif**) and includes it in the design files. You must always use "Project Save & Compile" each time you make memory changes using the DMDeditor.

e. Configure (program) the FLEX IC and observe the message sequence.

Method #2: Use the Altera "initialize memory" window.

NOTE: *If you have used the DMDeditor provided on the CD-ROM, you do not need to use method #2. You can proceed to the "Lab Exercise" section.*

The procedure is similar to the light sequencer system procedure for Lab 11. Here is the procedure:

a. You must begin by completing the "Project Save & Compile" step.

b. From the **MAX+PLUS II** menu select **Simulator**. The "Simulator" window will open. It is shown Figure 11-27 of Lab 11.

c. From the menu that runs across the very top of the screen, click on **Initialize** then select **Initialize Memory**. The "Initialize Memory" window shown in Figure 12-33 will open.

d. Enter the "Sample Program #1" memory map data from Figure 12-24. Start by clicking on the **000** entry on the first line beside address "00" and change it to the value in the memory map diagram. Press the **<tab>** key and change the next **000** entry to the right. Work your way across the first line of "000" numbers then proceed to the second line (address 08). Continue to make changes until all data from memory map diagram is entered. Sample Program #1 does not fill the entire memory initialization file.

e. Click the **Export File** button. Save the code you just entered as file **a:\labs\lab12.mif**. Click "OK."

f. You need to compile the project again. The compiler takes the MIF information (a:\labs\lab12.mif) and includes it in the design files. You must always use "Project Save & Compile" each time you make memory changes using the "Simulator."

g. Configure (program) the FLEX IC and observe the message sequence.

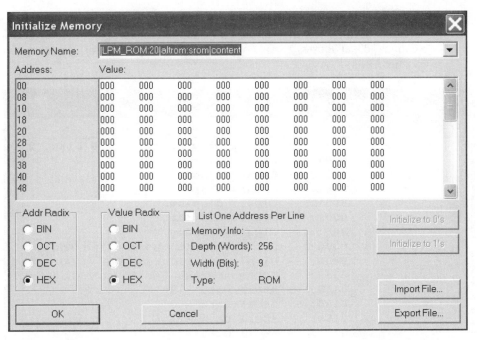

Figure 12-33 Initialize Memory Window

LAB EXERCISES AND QUESTIONS

This section contains lab exercises that can be performed on the UP board and questions that can be answered at home. Ask your instructor which exercises to perform and which questions to answer.

Lab Exercises

Exercise 1: Test the features of the DMDeditor.

You will use the DMDeditor to learn how to use its editing features. This lab exercise assumes you have used DMDeditor to enter "Sample Program #1" memory map diagram for Electronic Billboard System (see the "Lab Work Procedure" section).

   a. Use Windows Explorer to double-click on the **file a:\labs\DMDeditor.exe**. This will run the DMDeditor.

   b. Open the MIF file. From the **File** menu select **Open "lab12.mif"**. You will see the "Sample Program #1" code.

   c. The "√ Fix addresses and comments" command can be used to redocument code changes. Change the code at the end of line "address 1" from "0001" to "0002." The comment at the end of the line still reads **Clears display**. From the **File** menu select **√ Fix addresses and comments**. The DMDeditor will change the comment to **Returns cursor to home position**. This represents the function of the new code **0002**. Change the code at the end of line "address 1" back to **0001** and use **√ Fix addresses and comments** again. This will restore the change.

   d. The "√ Fix addresses and comments" command can be used to renumber the addresses from "0" to "FF" if lines have been moved, removed, or inserted. Remove the lines with code "0000" at addresses "1A" and "1B." You have created a gap in the address numbering. From the **File** menu select **Fix addresses and comments**. The DMDeditor will renumber all lines and fill the gap.

   e. The "√ Fix addresses and comments" command can be used to inform you that you have exceeded the maximum of 256 lines. If you copy a line and repaste it several times, it is conceivable that you will exceed the max line address of "FF" (or 255). When this occurs the white background will turn red. When you see a red background, you must find lines to delete and then reuse the "√ Fix addresses and comments" command. The background will turn back to white when you have deleted a sufficient number of lines. Test this feature now. Drag the mouse, mark the "Clear display command", at address 1, and copy it. Paste five copies of this command line. Refer to Figure 12-34.

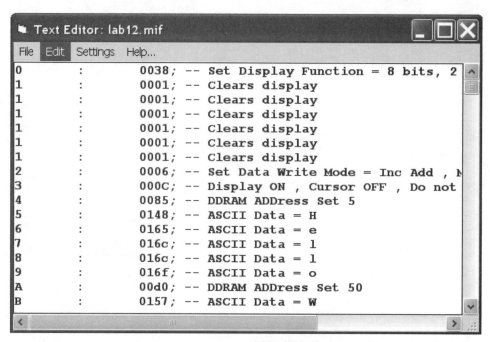

Figure 12-34 DMDeditor Window: Paste five copies of the clear display command.

From the file menu select √ **Fix addresses and comments.** You will see the addresses will be corrected (you won't have six entries for address 1) but the background will be red. Drag the mouse to select lines beyond address "1F" and delete three of them. From the file menu select √ **Fix addresses and comments.** You will continue to see the red background because too few lines were deleted. Select lines beyond the address "1F" and delete seven of them. From the file menu select √ **Fix addresses and comments.** This time, you will see the background turn back to a white color because more than enough lines were removed. There is no harm in removing more lines than necessary. This is useful if you have lost track of how many new lines have been inserted. Now remove the five extra "Clear display" commands, use √ **Fix addresses and comments,** and **Save & exit.**

Exercise 2: Electronic Billboard System Project Guidelines.
This lab exercise can be used as a design project.

Step A: Program the DMD to tell a story with basic level messaging.

Make up a story, slogan, or advertisement and display it on the DMD. Begin your story by displaying a couple of messages. You will add to the story in step B and step C. Detach and make copies of the programming form in Appendix D. Use the copies to create your DMD code. Include the following special effects while telling your story:

- With the cursor on, write a two-word message using two lines of the DMD. Write one word on the top line and a second word on the bottom line.
- With the cursor off, write a two- or three-word message off screen and scroll into view. It will replace the first message.
- With the cursor off, flash (blink) the message by turning the display on and off two times.

NOTE: *You can slow down the blink rate by sending two or three display on/off commands in succession.*

Step B: Program the DMD to continue the story with intermediate level messaging.

Continue with the story, slogan, or advertisement. You will add two messages to your story in this step of the lab. Do not delete the DMD code that you have created in step A. Add the step B code to the end of the step A code. You may want to use the "Clear Display" and "Send Cursor Home" commands to begin this section of new DMD code. Continue to fill in the programming forms you have detached and copied from Appendix D. Include the following special effects while telling your story:

- With the cursor off, write a multiple word message backwards (set I/D = 0) on the first line.
- With the cursor off, overwrite forward (set I/D = 1) only the second word of the message to change the meaning of the message. Example: Original message "The cat sleeps" changes to "The dog sleeps."
- With the cursor off, write a message on the second line so that characters appear in a random order. *Random* means characters appear out of their normal display sequence. The message won't be written consecutively one letter after the next from left to right or from right to left.
- Turn on the cursor and enable it to blink the character at the cursor position. Use the move cursor command to move the blinking cursor back and forth across the message on the second line.

NOTE: The system uses a 12 PPS clock rate. At 12 PPS the cursor may move over some characters so quickly that they will not blink.

Step C: Program the DMD to add CGRAM custom characters to the end of your story.

Use **CGRAM custom characters** to add an advanced special effect to your story. Be creative. Here are some ideas:

- Packman eating dashes
- A letter of the alphabet explodes
- An airplane towing a banner
- A happy face that smiles and winks

Review Sample Program #2 in the Introductory Information section of this lab, it will help you create custom characters. Appendix D contains a "CGRAM Drawing Guide" that you can detach, copy, and use to help you create custom characters. Continue to fill in the programming forms you have detached and copied from Appendix D. The limiting factor to your creativity may be the storage capacity of the ROM. Custom character initialization can use up a lot of the free address space in ROM. Work with the 256×9 storage capacity of the ROM. Do not increase the storage capacity of the ROM. Do not remove the DMD code for steps A and B. Let the size of the ROM be one of the challenges of this lab exercise. Be as creative as you can be using only a 256x9 ROM.

Exercise 3: Add an LCD digital display to the random number generator of Lab 5.

To complete this lab you need to connect a seven-segment LCD display to the FLEX expansion header. In Lab 5 you designed a random number generator system. The system displayed a random number between 0 and 9 when a pushbutton was pressed. In Lab 6 the number was displayed on the UP board digital LED display. In this lab you will use a seven-segment LCD display in place of the LED display.

You will need to use the information from Figure 12-5 in the Introductory Information section of this lab and the lab work procedure for Lab 5 to design the system. Unlike previous labs, you will not be given a "MAX+PLUS II" system diagram to follow as a guide. This will make the design a little more challenging.

Lab Questions

1. Make a copy of the programming form in Appendix D and use it to create a DMD program that will:
   - Display the first nine letters of your first name in the middle of the second line. The cursor must be visible.
   - Turn off the cursor and flash your name by turning the display on and off two times.
2. Make a copy of the programming form in Appendix D and use it to create a DMD program that will:
   - Write "Good day" backwards (set "I/D = 0") in the center of the first line. The cursor must be invisible.

- Overwrite forward (set "I/D = 1") only the word "day" and change it to "night." Do not rewrite the word "Good."

When you are done the message "Good night" will replace the message "Good day."

3. Make a copy of the programming form in Appendix D. Copy the following "HEX Address" and "HEX Value" DMD data onto the programming form.

	HEX Add.	HEX Value		HEX Add.	HEX Value		HEX Add.	HEX Value
1	00	038	13	0C	161	25	18	01C
2	01	001	14	0D	120	26	19	01C
3	02	006	15	0E	149	27	1A	01C
4	03	00C	16	0F	006	28	1B	01C
5	04	0CC	17	10	0A2	29	1C	01C
6	05	177	18	11	155	30	1D	008
7	06	168	19	12	120	31	1E	008
8	07	161	20	13	173	32	1F	00C
9	08	174	21	14	161			
10	09	004	22	15	179			
11	0A	083	23	16	120			
12	0B	16D	24	17	01C			

The DMD for this question has been connected to a 32×9 ROM. Analyze the code and figure out the effect it has on the DMD. Assume the code from address "00" is sent first, followed by address "01," and so on. Assume the DMD is being clock at "1 PPS" instead of "12 PPS." The DMD receives a new code every second.

a. Fill in the "Binary Value" and the "Description" on the programming form.

b. For each time interval shown in Figure 12-35, write a detailed description using "nontechnical language" of what the user will see on the DMD. Be sure to describe the cursor, the message, the direction the message is written, where the message is written, and what part of the message scrolls in and out of view. Each DDRAM table represents the viewable area on the DMD at various points in time through the ROM DMD code. Fill in each table to show what will appear on the viewable area of the two DMD lines at the end of each time interval.

CONCLUSIONS

- LCD technology is very energy efficient.
- LCD displays require a pulse waveform. DC destroys the nematic fluid.
- Dot matrix displays have pixels arranged in rows and columns.
- Dot matrix displays with the HD44780 controller are command driven.
- ASCII is a 7-bit code that is used to represent alphanumeric information.
- Character generator ROM is a nonvolatile memory system that stores the character pixel data.
- Display data RAM is a volatile memory system that stores the pixel data for each character that is currently being displayed.
- Character generator RAM is used to create custom characters.
- To use a DMD without reading the "busy flag," a ROM-based electronic billboard system must transfer codes to the DMD at a slower rate than the DMDs maximum code execution time.

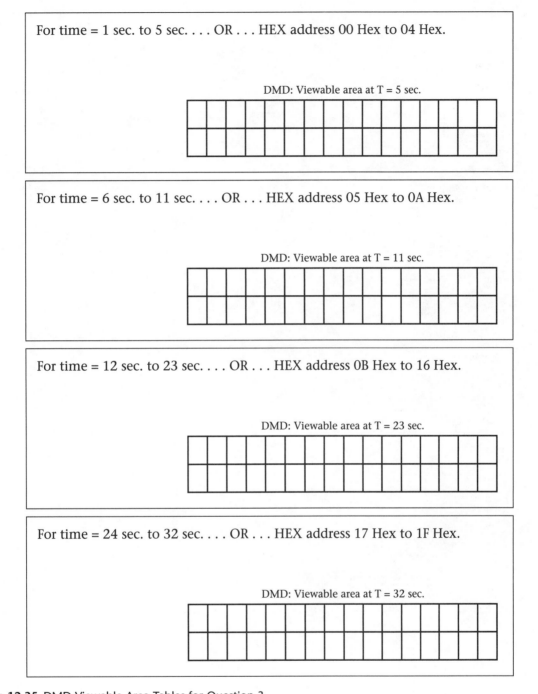

For time = 1 sec. to 5 sec. . . . OR . . . HEX address 00 Hex to 04 Hex.

DMD: Viewable area at T = 5 sec.

For time = 6 sec. to 11 sec. . . . OR . . . HEX address 05 Hex to 0A Hex.

DMD: Viewable area at T = 11 sec.

For time = 12 sec. to 23 sec. . . . OR . . . HEX address 0B Hex to 16 Hex.

DMD: Viewable area at T = 23 sec.

For time = 24 sec. to 32 sec. . . . OR . . . HEX address 17 Hex to 1F Hex.

DMD: Viewable area at T = 32 sec.

Figure 12-35 DMD Viewable Area Tables for Question 3

Appendix A

The Evolution of ROM and RAM

Types of ROM: MROM, PROM, EPROM, EEPROM, and FLASH

Lab 11 described the fundamental core of a ROM. Refer to Figure A-1. Figure A-2 adds detail to the fundamental core diagram. The memory cell is a **fuse** connected to the gate of a **FET**. The fuse connects the gate of the FET to the decoder output. Programming a ROM is a process that involves burning out a fuse to store "0" or leaving the fuse intact to store "1." When a fuse is burned out, the FET will be *off* because the gate lead can no longer be connected to "1." The *off* FET is connected to "0" through the resistor shown near the bottom of the diagram. When a fuse is left intact, the FET can be turned *on* by a "1" coming from the output of the "1 of 4"

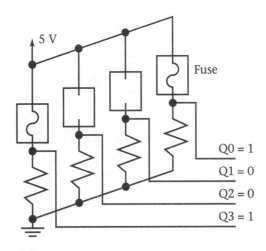

Figure A-1 4-bit ROM Fundamental Core

4x4 ROM

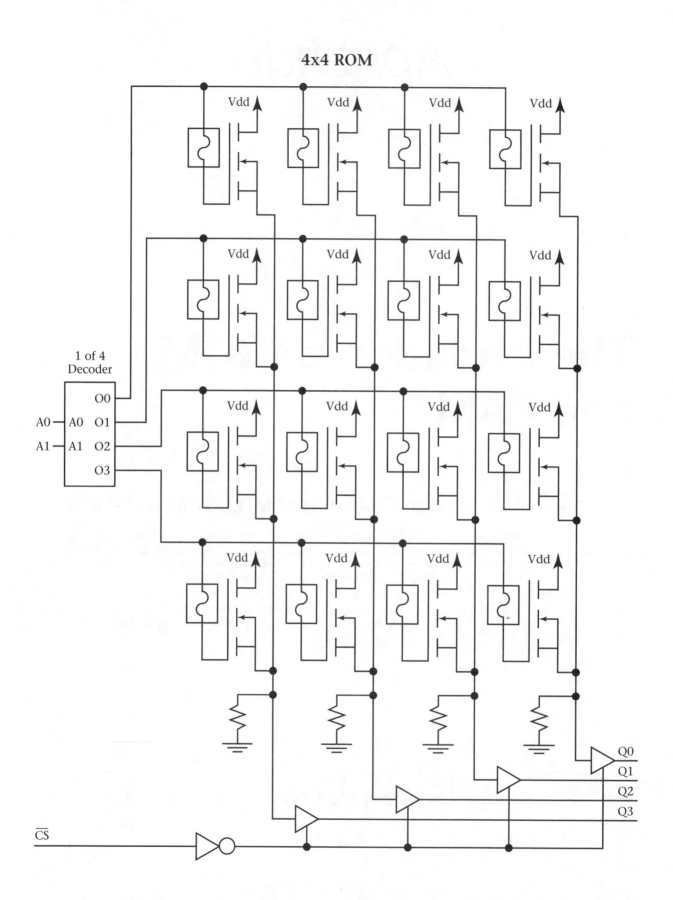

Figure A-2 4×4 ROM Architecture

decoder. A ROM programmer is a device that takes a blank ROM (all fuses intact) and uses a programming voltage (high voltage) to burn out fuses, which correspond to a "0" in memory. The programmer also can leave a fuse intact if the memory cell is to store a "1."

Reading a ROM is a simple process. **CS** is asserted with a "0." This allows the buffers to pass the data out and not to be in "HiZ" (review the information presented in Lab 11 if you are unsure what "HiZ" means). The address is applied to inputs "A1" and "A0." The "1 of 4" decoder will assert one of its outputs with a "1" and the other three outputs will stay at "0." The decoder output at "1" will energize a single row of four FETs. FETs with their fuses intact will turn *on*. FETs with burned fuses will stay *off*. *On* FETs will output a "1" to the data bus and *off* FETs will output a "0." The remaining three-decoder outputs at "0" will ensure that all other transistors are OFF regardless of the state of the FET fuse. This, in turn, prevents internal bus contention.

MROM is a *m*ask programmed ROM. As shown in Figure A-3, MROMs have no fuses. It is a precursor to fuse programmable technology. The manufacturer programs the MROM from a **memory map diagram** provided by the user. The memory map diagram indicates which gate leads to cut and which gate lead to leave intact. The MROM is then manufactured from a photographic mask. Figure A-3 shows a 4×4 MROM programmed with the numbers 9, 10, 11, and 12.

PROM technology uses **fuses**. It is a *p*rogrammable ROM that is mass-produced by the PROM manufacturer. Users purchase the affordable blank PROMs and a PROM programming station to program the PROMs themselves. The manufacturer is no longer required to do the programming. Figure A-2 shows a PROM.

EPROM technology uses a **UV (ultraviolet) erasable FET** in place of a fuse. EPROM is an *e*rasable and *p*rogrammable device. Placing a high-programming voltage on the device programs the EPROM and exposing it to UV light erases it. EPROMs have a UV window that is usually covered with a sticker to keep out stray UV light once they have been programmed.

EEPROM technology uses an **electrically erasable FET** in place of the UV erasable FET. EEPROM is an electrically erasable and programmable device. Placing a high-programming voltage on the device programs and erases the EEPROM. EEPROM eliminates the need to expose the device to UV light. EEPROMs are also referred to as E^{2}PROMs.

FLASH ROM is a technology that allows the ROM to be programmed and erased while the device is in the **target system**. The term *target system* refers to the electronic product sold to a consumer. MROM, PROM, and EPROM are all technology evolutions that required their programming to be done while the ROM is removed from the "target system." To program a ROM while it is in the "target system" requires a ROM that is capable of understanding FLASH ROM commands. It must be able to understand FLASH commands to program, verify, read, and erase the FLASH ROM. The target system connected to the FLASH ROM must have intelligence in order to be able to generate FLASH ROM commands and is typically a computer. The term *FLASH* refers to the ability of the ROM to be quickly erased and reprogrammed.

Figure A-4 shows the various types of ROMs. Figure A-4 (A) shows the MROM. Because MROMs are programmed by the manufacturer design, setup charges can be very expensive. Figure A-4 (B) is used to represent PROMs, EPROMs, and EEPROMS. The user programs all of these ROMs on site with the use of a ROM programmer. Design changes for PROM require the purchase of a new blank PROM because they are not erasable. Design changes for EPROMs and EEPROMs typically require the user to remove the ROMs from the target system in order that they be erased and reprogrammed. Before EPROMs can be reinserted into a ROM programmer, they must be erased by exposure to UV light. EEPROMs, on the other hand, are immediately reinserted into the ROM programmer, which performs both the erasing and reprogramming task. Figure A-4 (C) shows a FLASH ROM. FLASH ROM is a technology advancement that eliminates the need to remove the ROM from the target system to make ROM changes. The target system can accomplish all programming, erasing and reprogramming tasks. The target system must be able to send flash commands to the ROM and is typically a computer based device.

4x4 MROM

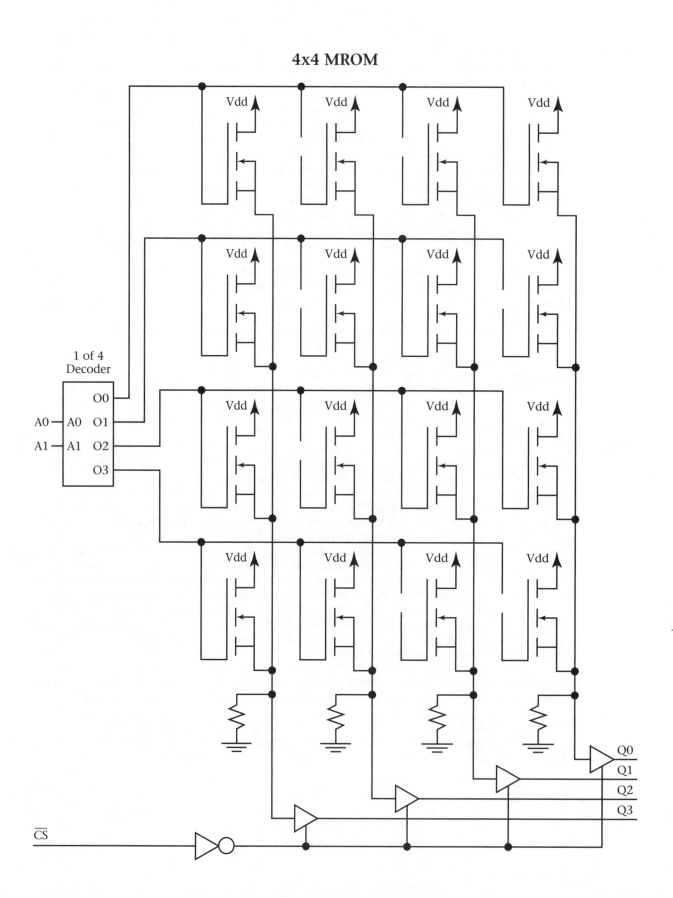

Figure A-3 4×4 MROM programmed with the numbers 9, 10, 11, and 12

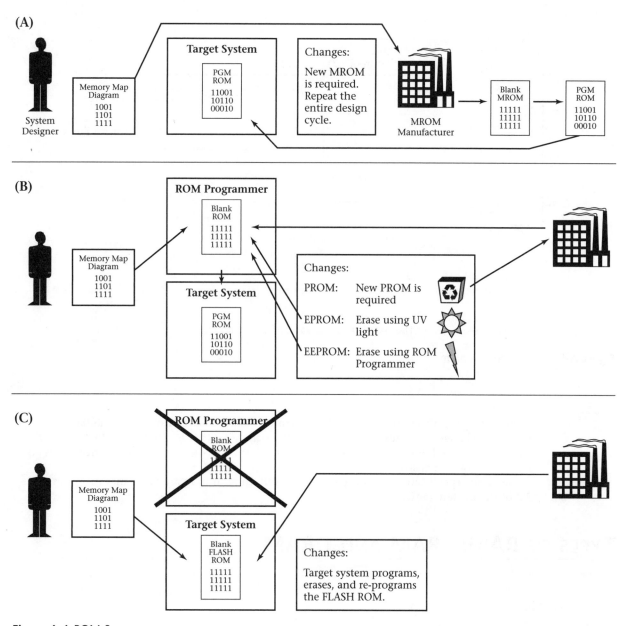

Figure A-4 ROM Summary

ROM TIMING DIAGRAMS

ROM data sheets show timing specifications using a waveform diagram. It is important to understand how to read these data sheets. Figure A-5 shows a timing diagram for a 2K×8 ROM.

The three buses of a ROM are each represented by a single waveform. The **address bus** waveform represents the 11 address inputs (A0 to A10). The dual horizontal lines represent that some address inputs are high while others are low. There is no intent on specifying the actual address. The "X" on the address bus, shown between the "old" and "new" address, represents a change of address.

The **Data Bus** waveform represents eight outputs (Q0, ... Q7). The single horizontal crosshatched line represents the data bus in "HiZ." The transition to two horizontal lines represent that some of the eight data bus outputs are high while others are low. There is no intent on specifying the actual data at outputs Q0 to Q7.

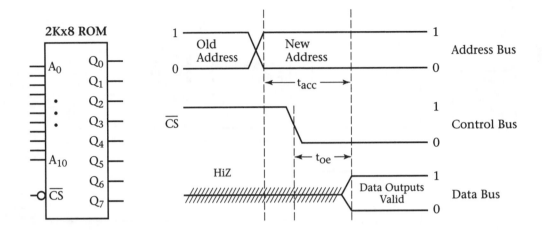

t_{oe} = Output Enable Table
Time required for the data bus to come out of HiZ

t_{acc} = Access Time
Time required for the data bus to be valid after the address has been changed

Figure A-5 ROM Timing Diagram

The timing specification t_{acc} is the **Access time**. It is a measure of how fast a ROM can respond to a change of address. "Access time" is a parameter that is often quoted when identifying the speed of memory systems. The smaller the number, the faster the memory is and, usually, the more expensive it costs.

The timing specification t_{oe} is the **output enable time**. It is a parameter that measures the speed of the tri-state buffers.

Types of RAM: SRAM and DRAM

SRAM

Figure A-6 was used in Lab 11 to explain RAM concepts. As seen in Lab 11 the fundamental core for RAM is a data register. A data register is made up of latches (or D flip-flops). RAM that is made up from flip-flops is called **static** RAM or **SRAM**. To build a **latch** using semiconductor requires several transistors. Simplifying the construction of the fundamental core increases the packing density of the semiconductor allowing more memory cells per square micron of silicone.

DRAM

A **dynamic** RAM or **DRAM** uses a capacitor as the fundamental core. A capacitor is a device that is structurally much simpler to fabricate and can easily be charged up to represent a "1" or discharged to represent a "0." The structural simplicity of DRAM makes it ideally suited for semiconductor miniaturization. Figure A-7 shows the storage cell for DRAM.

Storing data in DRAM is simple. Charge up the capacitor to store a "1" and discharge the capacitor to store a "0." The disadvantage of using a capacitor as a memory cell is the amount of time it takes for a charged capacitor voltage to decay, which, in turn, changes a binary "1" into a binary "0." A fully charged DRAM capacitor takes about 2 milliseconds to discharge. Unless the charges on the capacitors used to store binary "1" is replenished every 2 milliseconds, the entire DRAM memory system will appear to be storing all "0s." Replenishing the charge on the capacitors is a task called **DRAM refreshing**. Most DRAM systems incorporate both the storage

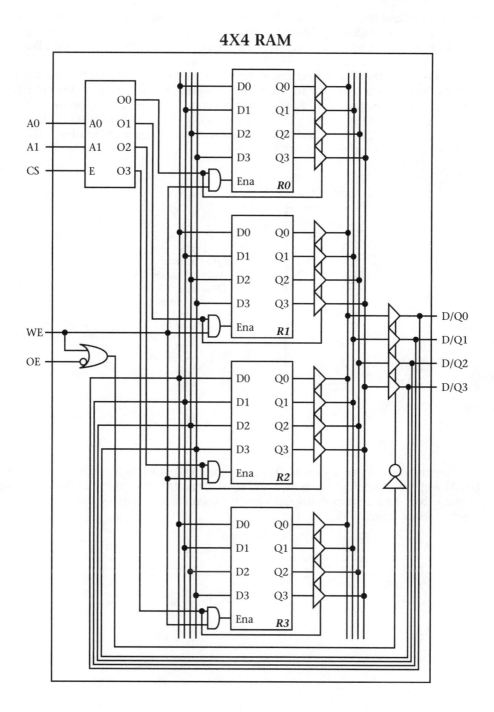

Figure A-6 4×4 Static RAM (SRAM)

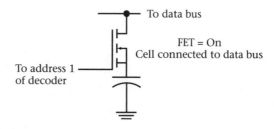

Figure A-7 Dynamic RAM (DRAM) Fundamental Core

operation task and the refresh task. The refreshing operation is not difficult to achieve because, in computer time, a 2-millisecond time interval is similar to a 2-month time interval for humans. DRAM systems can control data transfer and refresh the necessary memory cells before any data is lost.

DRAM technology was introduced at a time when creating a digital IC with a large quantity of inputs and outputs (I/O) was an engineering problem. Reducing the number of DRAM I/O was a high priority for engineers. DRAM technology simplified the construction of RAM so much so that it allowed engineers to create DRAM ICs with high levels of integration. High levels of memory integration, however, led to an IC that required a large number of I/O. Figure A-8 shows the number of I/O required by a 64K×1 DRAM.

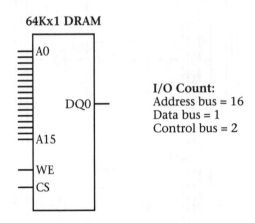

Figure A-8 64K×1 DRAM

Engineers found that they could reduce the number of I/O if they arranged the internal DRAM addressing architecture in rows and columns and then send the address bus information in two halves. **Address bus multiplexing** is the term used to describe this architecture. Refer to Figure A-9 and Figure A-10.

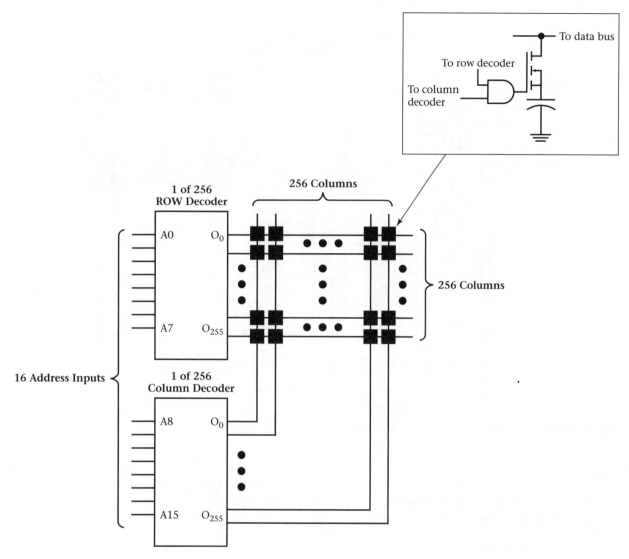

Figure A-9 64K×1 DRAM: Matrix layout

The number of address inputs has been reduced from "16" to "8" inputs; however, two address multiplexing control inputs, called **RAS** and **CAS**, are required. **RAS** (pronounced *raz*) is the **row address strobe** (or select) input. When the lower 8-bit row address is sent to the DRAM, RAS is asserted to store the **row address** in the data register. The address inputs can then be used to transfer the upper 8-bit column address. **CAS** (pronounced *caz*) is the **column address strobe** (or select). It is asserted to store the **column address** in the data register. The advantage of address bus multiplexing is the reduction of address inputs. The disadvantage is that it takes two 8-bit transfers to send a 16-bit address to the DRAM system. Figure A-11 shows an address bus multiplexing timing diagram.

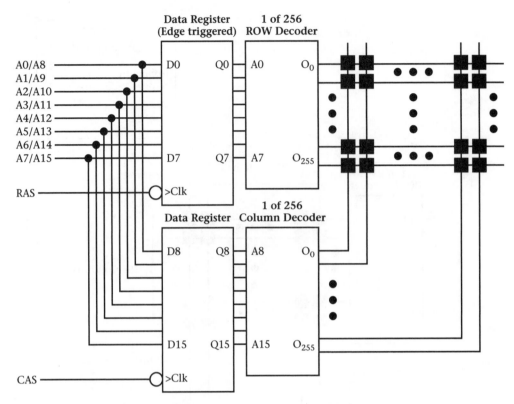

Figure A-10 64K×1 DRAM: Address Bus Multiplexing

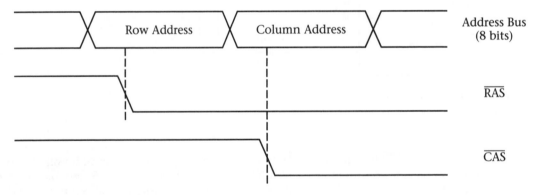

Figure A-11 Address Bus Multiplexing Timing Diagram

DRAM Evolution

Over the years, steps to improve the operating speed of DRAM have been necessary in order to keep up with faster and faster computer systems. A brief description of DRAM technology improvements follows.

Fast Page Mode (FPM) DRAM

Whereas standard DRAM requires that a row address and column address be sent for each access, **FPM** works by sending the row address just once and then sending a series of column addresses. Sending new column addresses accesses the data on a page (on the same row).

Extended Data Out (EDO)

EDO memory can latch the data on its output pins so that a new address can be sent while the data from the previous address is being read by the computer system. EDO improvements are marginal at best.

Synchronous DRAM (SDRAM)

SDRAMs can refresh memory cells internally. SDRAMs can transfer entire blocks of data once they are sent the starting address. This type of transfer is called **burst transfer**. Burst transfers are fast because they synchronize the DRAM to a computer's system clock instead of the CAS control line. SDRAMs have two internal memory banks. One memory bank can be completing a transfer while the other is setting up for the next transfer. This two-bank system is called memory interleave.

Double Data Rate Synchronous DRAM (DDR SDRAM)

SDRAM transfers data on a single edge of the bus system clock. **DDR SDRAM** allows data transfers on both the positive and negative edges of the bus system clock. This doubles the transfer rate without increasing the bus system clock pulse rate.

Direct Rambus DRAM (DRDRAM)

DRDRAM is a proprietary technology proposed by Rambus Inc. in partnership with Intel. DRDRAM is based on the high-speed Direct Rambus Channel. It is a 16-bit bus running at a speed of 800 MHz.

Figure A-12 shows the dates and speed limits of the various DRAM catagories.

Year Introduced	Technology	Speed Limit
1987	FPM	16–66 MHz
1995	EDO	33–75 MHz
1997	PC66 SDRAM	66 MHz
1998	PC100 SDRAM	100 MHz
1999	RDRAM	800 MHz
1999/2000	PC133 SDRAM	133 MHz
2000	DDR SDRAM	266–400 MHz
2003/2004	DDR-2 SDRAM	>400 MHz

Figure A-12 DRAM Evolution

MEMORY SYSTEM DESIGN

Many memory systems are built up by combining smaller memory systems together. For example, how would you design a 16×8 ROM memory system from 16×4 memory ROMs? How would you design a 32×4 ROM memory system from 16×4 ROMs? This section of the lab book studies the techniques used to design memory systems.

Data Bus Expansion Design: Combine 16×4 ROMs to Make a 16×8 ROM System

In this first case, the data bus of the system is larger than the data bus of a single ROM. This type of expansion is called **data bus expansion**. The first step is to use the system storage capacity equation to draw the address bus lines at the top and the data bus lines at the bottom. The system storage capacity equation is **16×8**. Refer to Figure A-13.

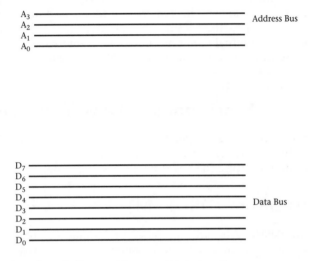

Figure A-13 Data bus expansion: Step 1: Draw address and data bus lines.

The second step is to figure out the number of ROMs required.

$$\text{Number of ROMs} = \frac{\text{System storage capacity equation}}{\text{Single ROM storage capacity equation}} = \frac{16 \times 8}{16 \times 4} = 1 \times 2 = 1 \text{ bank of 2 ROMs}$$

The term **1 bank of 2 ROMs** implies that you would group two ROMs together to create a **bank** that would have a data bus large enough to satisfy the design requirement. All ROMs in the bank are drawn in a three-dimensional plane behind each other. Refer to Figure A-14.

The 3-D diagram implies that the front ROM has four address inputs A3, A2, A1, and A0 and that they are connected to the back ROMs address inputs. The 3D diagram also implies that the back ROM has a chip select (CS) input and that it is connected to the front ROM. The last step is to connect the ROMs to the buses. Refer to Figure A-15.

The final result is a memory system with an 8-bit data bus that is derived from ROMs with 4-bit data buses.

System Memory Map Diagram (SMMD)

A memory map diagram is a block diagram that shows the address range and data bus assignment for each ROM in the system. A general layout for a SMMD is shown in Figure A-16.

To generate the SMMD for the 16×8 memory system, draw a box with a vertical split to represent the data bus expansion. The label at the top of the box shows the data bus split, and the unit numbers are placed inside the box. Refer to Figure A-17.

The SMMD clearly shows which ROM controls the upper section of the data bus D7...D4 and which ROM controls the lower section of the data bus D3...D0. To complete the SMMD, the starting and last hexadecimal addresses are added. Writing the address in binary and then converting it to HEX will prevent you from making mistakes. Refer to Figure A-18.

The final state of the SMMD shows how each ROM is addressed and which section of the data bus it is controlling.

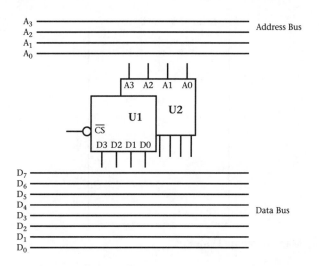

Figure A-14 Data bus expansion: Step 2: Place ROMs between the bus lines.

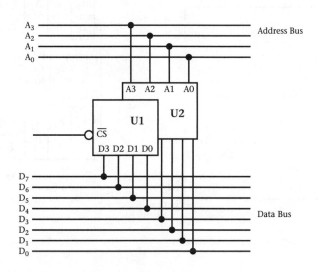

Figure A-15 Data bus expansion: Step 3: Connect ROMs to buses.

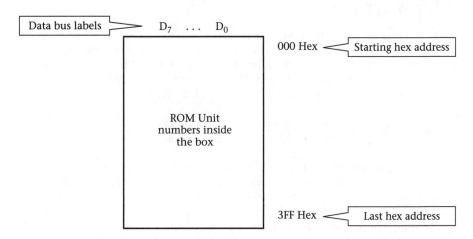

Figure A-16 System Memory Map Diagram: General Layout

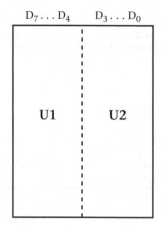

Figure A-17 System Memory Map Diagram: Data Bus

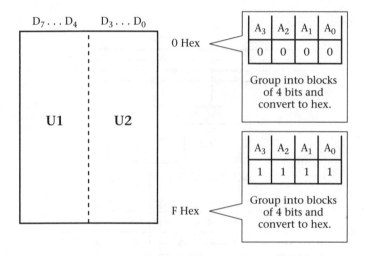

Figure A-18 System Memory Map Diagram: Addresses

Address Bus Expansion Design: Combine 16×4 ROMs to Make a 32×4 ROM System

In this second case, the address bus of the system is larger than the address bus of a single ROM. This type of expansion is called **address bus expansion**. The first step is to use the system storage capacity equation to draw the address bus lines at the top and the data bus lines at the bottom. The system storage capacity equation is 32×4. Refer to Figure A-19.

The second step is to figure out the number of ROMs required.

$$\text{Number of ROMs} = \frac{\text{System storage capacity equation}}{\text{Single ROM storage capacity equation}} = \frac{32 \times 4}{16 \times 4} = 2 \times 1 = 2 \text{ banks of 1 ROM}$$

The term **2 banks of 1 ROM** implies that each ROM creates a storage bank of 16 4-bit numbers. All ROMs in the bank are drawn side by side. Refer to Figure A-20.

The next step is to connect the ROMs to the buses. Refer to Figure A-21.

As can be seen in Figure A-21, both U1 and U2 share the data bus lines D3, D2, D1, and D0. To prevent bus contention, the extra address input "A4" will be connected to a "1 of X" decoder. U1 is selected when "A4" is low and U2 is selected when the "A4" is high. The memory system will store half the numbers in U1 when "A4" is low and the second half in U2 when "A4" is high. Refer to Figure A-22.

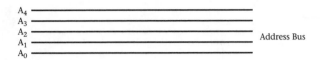

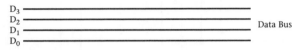

Figure A-19 Address bus expansion: Step 1: Draw address and data bus lines.

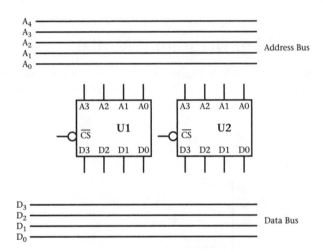

Figure A-20 Address bus expansion: Step 2: Place ROMs between the bus lines.

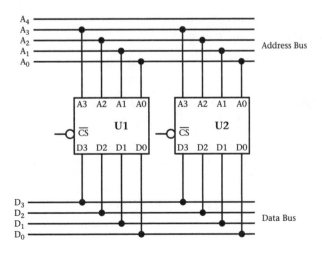

Figure A-21 Address bus expansion: Step 3: Connect ROMs to the buses.

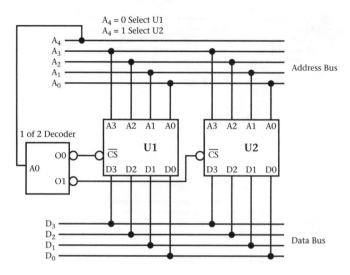

Figure A-22 Address bus expansion: Step 4: Connect "1 of X" decoder.

System Memory Map Diagram (SMMD)

To generate an SMMD for the 32×4 memory system, you must draw a box with a horizontal split to represent the address bus expansion. The label at the top of the box shows the data bus, and the unit numbers are placed inside the box. Refer to Figure A-23.

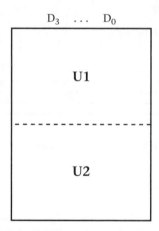

Figure A-23 System Memory Map Diagram: Data Bus

The HEX starting and last addresses need to be added. Writing the address in binary and then converting it to HEX will prevent you from making mistakes. Refer to Figure A-24.

The final step is to add the last address of U1 and the starting address of U2. The last address of U1 has all the address inputs of U1 at 1 while the decoder asserts the chip select input. The first address in U2 has all the address inputs of U2 at 0 while the decoder asserts the chip select input. Refer to Figure A-25.

The final state of the memory map diagram shows the address range occupied by each ROM and that each ROM is connected to the entire data bus.

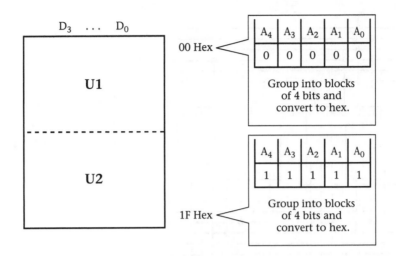

Figure A-24 System Memory Map Diagram: System Starting Address and Last Address

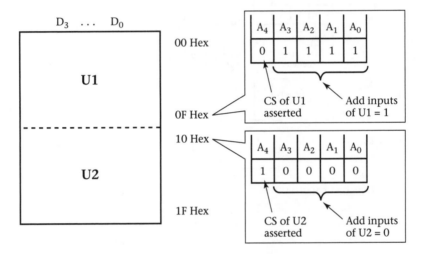

Figure A-25 System Memory Map Diagram: Starting and Last Address of U1 and U2

Data and Address Bus Expansion Design: Combine 32×4 ROMs to Make a 64×8 ROM System

In this final case, the address bus and the data bus of the system are larger than the address bus and data bus of a single ROM. The procedure used to design the system combines the steps described in the previous two sections "Address Bus Expansion Design" and "Data Bus Expansion Design." A summary of the steps used to design the system is shown in Figure A-26.

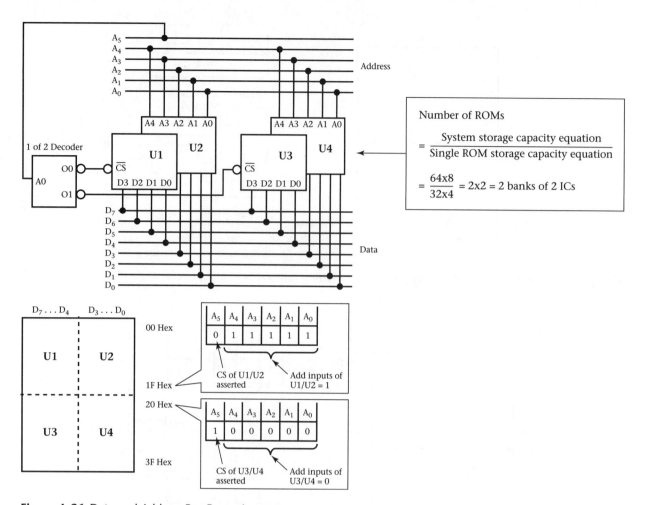

Figure A-26 Data and Address Bus Expansion

QUESTIONS

1. Study the memory system in Figure A-27.
   a. What is the storage capacity equation for U3?
   b. What is the storage capacity equation for the entire memory system?
   c. Draw the memory map diagram for the system.
2. Draw a diagram for a 256K×8 memory system that uses RAMs. Each RAM has a storage capacity of 64K×4 and two active low chip selects.
   a. Connect the two active low chip selects together and design the system with a "1 of X" decoder connected to the extra system address lines.
   b. A RAM with two chip selects must have both chip selects asserted or else the data bus lines will be in high impedance (HiZ). Because each RAM has two chip selects, the "1 of X" decoder can be replaced by 2 NOT gates. Try and figure out how to connect the 2 NOT gates to replace the "1 of X" decoder. Do not connect the two active low chip selects together.

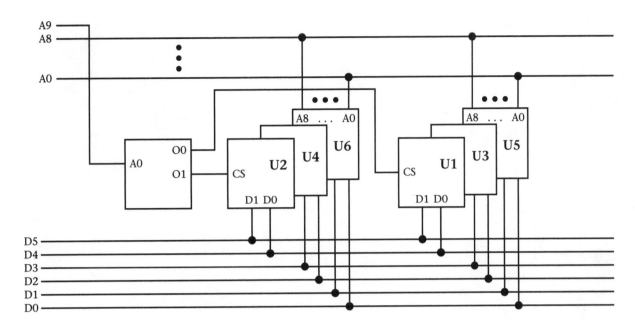

Figure A-27 Memory System Diagram for Question #1

Appendix B

VHDL Design Guide

CORRECTING VHDL SYNTAX ERRORS

Syntax errors are like English grammar errors. Altera needs each line of VHDL code to have the correct punctuation. Here is what happens when you use "Create a Default Symbol" on the "vending.vhd" file that has a "," missing at the end of one line. Refer to Figure B-1.

If you scroll up to the first error message in the window and double-click on it, the Altera software will show you the location of the error. The cursor will appear in the VHDL text editor window on the line after the error. Refer to Figure B-2.

The error is usually located on the line above the line with the cursor in it. The line with the error creates a problem in the next line. The next line and many others are wrongfully identified as having problems. Always fix the first error shown in the error message window, then retry **Create a Default Symbol.** One fix usually clears many, if not all, the errors.

```vhdl
LIBRARY ieee;
USE ieee.STD_LOGIC_1164.ALL;
ENTITY vending IS
        PORT(
                Q1, Q2, Q3, L              : IN    STD_LOGIC;
                P, C                       : OUT   STD_LOGIC);
END vending;

ARCHITECTURE a OF vending IS
        SIGNAL input: STD_LOGIC_VECTOR (3 DOWNTO 0);
        SIGNAL output: STD_LOGIC_VECTOR (1 DOWNTO 0);
BEGIN
-- Concurrent Signal Assignment
 input (3) <= L;
 input (2) <= Q3;
 input (1) <= Q2;
 input (0) <= Q1;

-- Selected Signal Assignment
 WITH input SELECT
        output <=       "00" WHEN "0000",
                        "00" WHEN "0001",
                        "00" WHEN "0010",
                        "00" WHEN "0011",

                        "00" WHEN "0100",
                        "00" WHEN "0101",
                        "00" WHEN "0110",
                        "10" WHEN "0111",

                        "11" WHEN "1000",
                        "11" WHEN "1001",
                        "11" WHEN "1010",
                        "11" WHEN "1011",

                        "11" WHEN "1100",
                        "11" WHEN "1101",
                        "11" WHEN "1110",
                        "11" WHEN "1111",

                        "00" WHEN others;

 P      <=      output(1);
 C      <=      output(0);
END a;
```

Remove this "," and get many error messages.

Click to scroll up to the first error message.

MAX+plus II - Compiler

⚠ Symbol generation was unsuccessful
9 errors
0 warnings

[OK]

Messages - Compiler

Error: Line 29: File a:\labs\vending.vhd: VHDL syntax error: selected signal assignment must have ';', but found <string_literal> "10" instead

Error: Line 20: File a:\labs\vending.vhd: VHDL syntax error: expected choices in selected signal assignment statement

◄ Message ► 0 of 10   ☐ Locate in Floorplan Editor   Help on Message

◄ Locate ► 0 of 0   Locate All

Figure B-1 VHDL Code with Errors

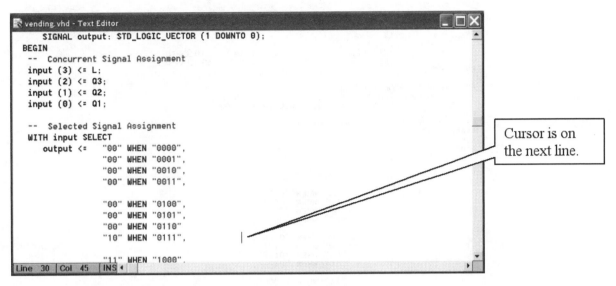

Figure B-2 Altera "Text Editor" Window Showing Error Location

Making Changes to a Functional VHDL Design

To build a VHDL system you start with the text editor and enter the VHDL code. You then use "create a default symbol" to look for syntax errors and create the symbol. Next you enter the VHDL symbol on the graphics editor page. Finally, you place input/output symbols on the page, save and compile, and program the CPLD to test the design. If the system does not work, you will need to make changes to the VHDL code. There may not be any syntax errors but the system VHDL code needs to be changed. Here is the procedure that you can use to change Lab 2:

1. View the design using the Graphic Editor. Refer to Figure B-3.

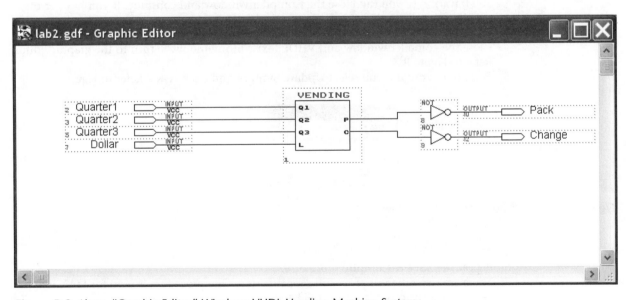

Figure B-3 Altera "Graphic Editor" Window: VHDL Vending Machine System

2. Double-click on the VHDL symbol **vending**. The text editor will open and display the VHDL code. Refer to Figure B-4.

```
LIBRARY ieee;
USE ieee.STD_LOGIC_1164.ALL;
ENTITY vending IS
        PORT(
                Q1, Q2, Q3, L          : IN    STD_LOGIC;
                P, C                   : OUT   STD_LOGIC);
END vending;

ARCHITECTURE a OF vending IS
        SIGNAL input: STD_LOGIC_VECTOR (3 DOWNTO 0);
        SIGNAL output: STD_LOGIC_VECTOR (1 DOWNTO 0);
BEGIN
  -- Concurrent Signal Assignment
  input (3) <= L;
  input (2) <= Q3;
  input (1) <= Q2;
  input (0) <= Q1;

  -- Selected Signal Assignment
  WITH input SELECT
        output <=         "00" WHEN "0000",
                          "00" WHEN "0001",
                          "00" WHEN "0010",
                          "00" WHEN "0011",

                          "00" WHEN "0100",
                          "00" WHEN "0101",
                          "00" WHEN "0110",
                          "10" WHEN "0111",

                          "11" WHEN "1000",
                          "11" WHEN "1001",
                          "11" WHEN "1010",
                          "11" WHEN "1011",

                          "11" WHEN "1100",
                          "11" WHEN "1101",
                          "11" WHEN "1110",
                          "11" WHEN "1111",

                          "00" WHEN others;

  P       <=    output(1);
  C       <=    output(0);
END a;
```

Figure B-4 Altera "Text Editor" Window with VHDL Code

3. Study the code and make the necessary changes. Click on the **File** menu and select **Create Default Symbol**. The Compiler window will open and run. If you get 0 errors and 0 warnings, you can close the Compiler window and continue. If you have errors, use the previous section titled "Correcting VHDL Syntax Errors" to correct errors. Continue from this point when all errors have been corrected.
4. Close the compiler window and VHDL text editor window. Return to the graphic editor. Refer to Figure B-3.
5. From the **Symbol** menu select **Update Symbol** and click "OK." Refer to Figure B-5.

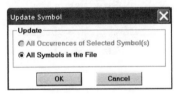

Figure B-5 Altera "Update Symbol" Window

6. Repeat the **Save & Compile** procedure.
7. There are two different procedures for this step. One procedure is used for MAX IC designs and the other is used for FLEX IC designs.

For a MAX IC Design: Reopen the **report file**, and check for pin number changes. Every time you compile, the pin numbers may change. You must check the numbers and rewire switches and LEDs if necessary.

For a FLEX IC Design: Pins used to connect the onboard switches and LEDs have been assigned to each input and output symbol. There is no need to take any action for this step.

These steps will ensure that the latest revisions to the VHDL code are programmed into the CPLD.

Appendix  C

FLEX Expansion Header Guide

FLEX EXPANSION HEADER PIN NUMBER DIAGRAM

The UP board has three 60-pin connectors that surround the 240-pin FLEX10K20 (or UP-2 FLEX10K70). They are labeled **Flex Expansion Header** "A," "B," and "C." A 60-pin socket must be soldered to the UP board Flex Expansion Header "A" in order to complete labs that require wire connections to a keypad or a digital display. Labs that connect two UP boards together also require wiring access to Altera FLEX IC. UP boards are shipped from the factory with wire headers on the MAX IC but not on the FLEX IC. Refer to Figure C-1. The pin layout diagram for FLEX Expansion Header A is shown in Figure C-2.

The diagram shown in Figure C-2 can be used to easily locate the pin numbers on the FLEX Expansion header socket. Each pin number is shown in a box along with a numbered counting guide (1...to...15). The counting guide shows you the number of pins you must count off from each end of the header socket in order to make a wire connection.

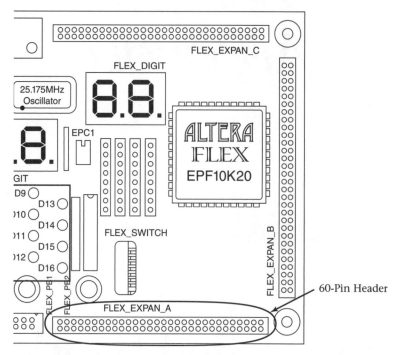

Figure C-1 FLEX Expansion Header

FLEX Expansion Header A

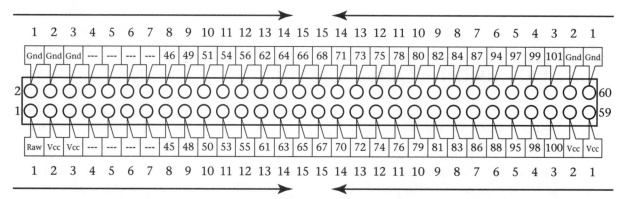

Figure C-2 FLEX Expansion Header "A" Pin Numbers

CONNECTING A DMD TO THE FLEX EXPANSION HEADER

The DMD can be connected to the FLEX expansion header several ways. A set of wires can be soldered to the DMD and the wires inserted into the "FLEX Expansion A" header. Refer to Figure C-3.

Wire wrap pins can be soldered to the DMD. Wire wrap pins are solid rectangular posts that would allow the DMD to be mounted directly onto the FLEX header. Refer to Figure C-4.

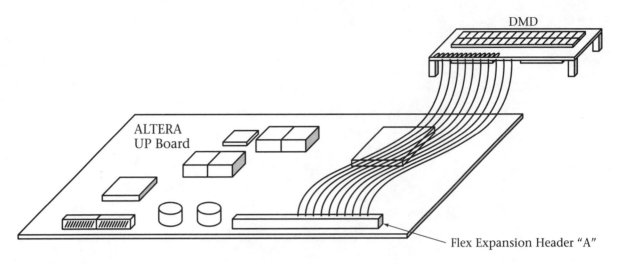

Figure C-3 DMD with Wire Connections to FLEX Header

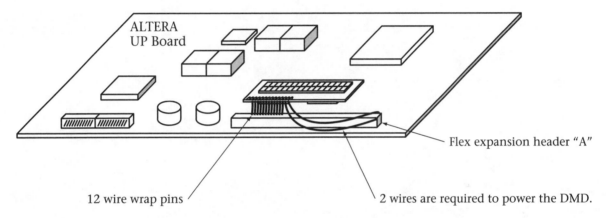

Figure C-4 DMD with wire wrap pins mounted directly to FLEX Header

Appendix **D**

Forms and Guides for the DMD Lab 12 Project

DMD COMMAND SUMMARY

COMMAND	R/W	RS	DB7	DB6	DB5	DB4	DB3	DB2	DB1	DB0	DESCRIPTION
CLEAR DISPLAY	0	0	0	0	0	0	0	0	0	1	Clears display and returns cursor to home position (DDRAM address = 00H).
SEND CURSOR HOME	0	0	0	0	0	0	0	0	1	*	Returns cursor to home position. Returns a shifted display to original position. Data in DDRAM is not changed.
SET THE DATA WRITE MODE	0	0	0	0	0	0	0	1	I/D	S	I/D sets the CURSOR/DISPLAY move direction. The CURSOR/DISPLAY will automatically move to the next DDRAM/CGRAM address after each data write. S specifies to shift the display or to move the cursor.
SET DISPLAY/ CURSOR ON/OFF	0	0	0	0	0	0	1	D	C	B	D sets display On/Off. C sets the cursor On/Off. B enables the character at the cursor position to Blink/Not Blink.
SHIFT/MOVE CURSOR/DISPLAY	0	0	0	0	0	1	S/C	R/L	*	*	S/C moves the cursor or shifts the display. Contents of DDRAM are not changed. R/L chooses the direction of the Shift/Move.
SET DISPLAY MODE (DISPLAY FUNCTION)	0	0	0	0	1	DL	N	F	*	*	DL sets the Data Bus size. N sets the number of lines to be used. F sets the Dot Matrix font (first instruction).
SET CGRAM ADDRESS	0	0	0	1	ACG (6-bit address)						Sets the starting CGRAM address. The CGRAM data is sent to this starting address after this setting.
SET DDRAM ADDRESS	0	0	1	ADD (7-bit address)							Sets the starting DDRAM address. The DDRAM ASCII/Japan/Custom characters are sent to this starting address after this setting.
READY BUSY FLAG/ADDRESS	1	0	BF	AC							Read the Busy Flag (BF) indicating internal operation is being performed and reads address counter contents.
CGRAM/DDRAM DATA WRITE	0	1	8-bit code for ASCII/Japan/Custom Characters								Writes data into DDRAM or CGRAM. Used after address has been set. The address pointer is inc/dec automatically.
READ DDRAM/ CGRAM DATA	1	1	Read Data								Read data from DDRAM or CGRAM.

I/D = 1: Increment the DDRAM address.
     0: Decrement the DDRAM address.

S = 1: Shift the display after data write. Cursor freeze.
    0: Move the cursor after data write. Display freeze.

D = 1: Turn the entire display ON.
    0: Turn the entire display OFF.

C = 1: Turn the cursor ON.
    0: Turn the cursor OFF.

B = 1: Blink the character at the cursor position.
    0: DO NOT blink the character at the cursor position.

S/C = 1: Shift the display.
     0: Move the cursor.

R/L = 1: Shift/Move to the RIGHT.
     0: Shift/Move to the LEFT.

DL = 1: Use an 8-bit data bus.
     0: Use a 4-bit data bus.

N = 1: Use 2 lines of the display.
    0: Use 1 line of the display.

F = 1: 5x10 dot matrix font.
    0: 5x7 dot matrix font.

BF = 1: Busy doing an internal operation.
     0: Ready to communicate.

DDRAM: Display Data RAM.
CGRAM: Character Generator RAM.

ACG: CGRAM Address
ADD: DDRAM Address

AC: CGRAM/DDRAM Address Counter
*: Not used. Set it to 0.

Figure D-1 DMD Command Summary

CGROM/CGRAM TABLE

6-Bit Addresses For CGRAM Initialization
CG0 address 00 0000
CG1 address 00 1000
CG2 address 01 0000
CG3 address 01 1000
CG4 address 10 0000
CG5 address 10 1000
CG6 address 11 0000
CG7 address 11 1000

Figure D-2 CGROM/CGRAM Table

DMD Programming Form (or Memory Map Diagram)

HEX Address	HEX Value	Binary Value									Description
		RS	DB7	DB6	DB5	DB4	DB3	DB2	DB1	DB0	

Figure D-3 Programming Form

DMD CGRAM DRAWING GUIDE

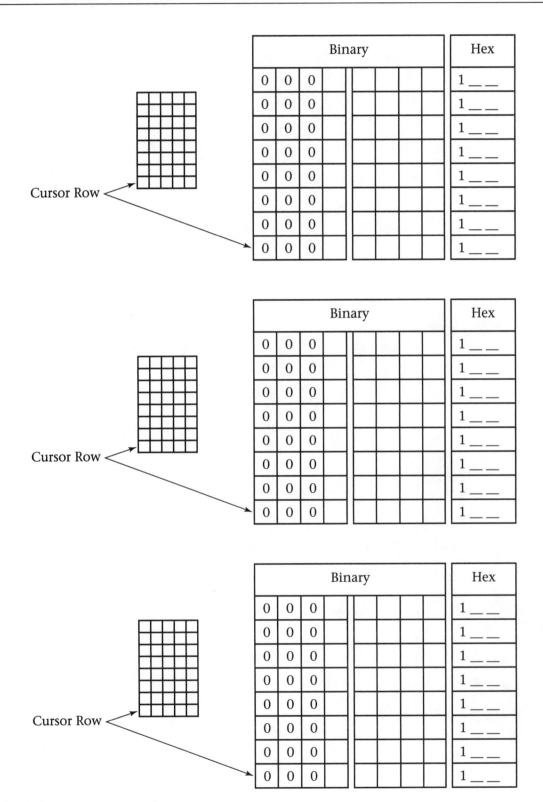

Figure D-4 CGRAM Drawing Guide

Appendix

Summary Sheet for FLEX IC Designs: Using "Lab 1" as a Guide

1 and 2. Turn on the computer (PC). Insert a blank formatted floppy disk into drive A:. Use Windows Explorer to quick format the floppy. This will erase all files. Use Windows Explorer to create a folder (directory) on the floppy called **Labs**.

3. Create a new project file.
   a. Start the **MAX+PLUS II** program.
   b. Create a project file: Click the **File** menu and select **New**. Select **Graphic Editor File** (or Text Editor File for VHDL designs) and click the "OK" button.
   c. Save the project. Click the **File** menu, select **Project**, and then **Set Project to Current File**. Replace the word **Untitled2.gdf** with **Lab1.gdf**. Select "A:" from the **DRIVES:** drop down menu and then select directory (folder) **Labs** (double-click on the folder name). Click the "OK" button.

4 and 5. Enter symbols.
   a. With the selection pointer (pointer you can move with the mouse), double-click in an empty space in the Graphic Editor window. You will see the Enter Symbol window.
   b. Double-click on a **library name**. This will display the symbols in the selected library in the Symbol Files box. Locate and double-click on the symbol from the menu. The symbol will be placed on the Graphic Editor worksheet.

6 and 7. Enter INPUT and OUTPUT symbols. Double-click the selection pointer on the drawing to open the Enter Symbol window. Double-click on the **c:\maxplus2\max2lib\prim** library. Locate and double-click on the INPUT or OUTPUT symbol from the menu.

8 and 9. Name the INPUT and OUTPUT symbols. Double-click the selection pointer on the word **PIN_NAME** of the input or output symbol and change it.

10. Draw lines (wires). From the **OPTIONS** menu select **Line Styles** and then the **Solid Line** (option at the top of the window). Place the selection pointer on the line at the

end of the input symbol. The Selection Pointer turns into a "+" shape drawing tool. While pressing the mouse button, drag the mouse to draw the line.

NOTE 1: *A green dotted box surrounds each symbol.*
This is a symbol boundary box. You must never run wires through (inside) the boundary box. You must always terminate connections at the stub end of a symbol's boundary box. Wires that run on top of a boundary box line, on the input or output stub, will automatically be connected to the symbol. You must provide a space between the wire and the symbol boundary box line when you run a wire pass a symbol.

NOTE 2: *To anchor a bend (or elbow) in the wire.*
You must release the mouse button and reclick it. A nonanchored elbow will move around as you move the mouse.

11$_{\text{FLEX}}$. Select the FLEX IC. The UP Board has two ICs. Your design will be programmed into the FLEX IC. From the **Assign** menu select **Device**. Uncheck **Show only fastest speed grades** and then select **Device Family FLEX10K**, and **Devices EPF10K20RC240-4** (for the UP-2 board, use EPF10K70RC240-4.) Click the "OK" button.

12$_{\text{FLEX}}$. Check your worksheet for basic errors.
a. From the **File** menu select **Project Save & Check**. If the Project Save & Check menu item is grayed out (not selectable), you will need to select **Set Project to Current File** and try again to select the **Project Save & Check** menu item.
b. If the Compiler issues any error messages, you should double-click on the first error message in the list. The MAX+PLUS II software will revert to the Graphic Editor and show the location of the error in *red*. Study the diagram and correct the problem. Repeat the Save & Check procedure.

13$_{\text{FLEX}}$. Assign switches to input of system.
Printed circuit board wire connections (PCB traces) connect eight switches to the FLEX IC.
To connect the switches, you must assign a pin number to each input symbol. Place the selection pointer over an input symbol. Right click and select **Assign** then **Pin/Location/Chip**....
Refer to Figure E-1.

FLEX Switch Pin Assignments

	EPF10K Pin #	
FLEX_SWITCH-1	41	Switch closest to the bottom of the UP board
FLEX_SWITCH-2	40	
FLEX_SWITCH-3	39	
FLEX_SWITCH-4	38	
FLEX_SWITCH-5	36	
FLEX_SWITCH-6	35	
FLEX_SWITCH-7	34	
FLEX_SWITCH-8	33	Switch closest to the top of the UP board
FLEX_PB1	28	Pushbutton switch on the left side of the UP board
FLEX_PB2	29	Pushbutton switch on the right side of the UP board

Oscillator = pin #91

Figure E-1 Altera UP Board FLEX Switch Pin Assignments

14$_{\text{FLEX}}$. Assign LEDs to the output of the system. PCB traces connect 16 LEDs to the FLEX IC. You can use the LEDS as a two-digit digital display or you can use D1 ... D8 as two groups of four individual bar LEDs.

a. To connect the LEDs you must assign a pin number to each output symbol. Place the selection pointer over the output symbol, right click, and select **Assign** then **Pin/Location/Chip...**. Refer to Figure E-2.

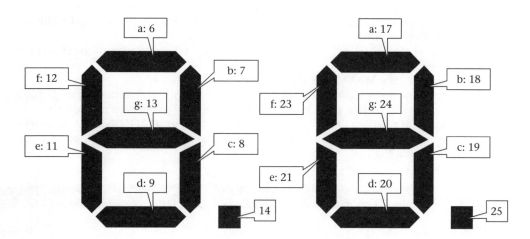

FLEX Digital Display

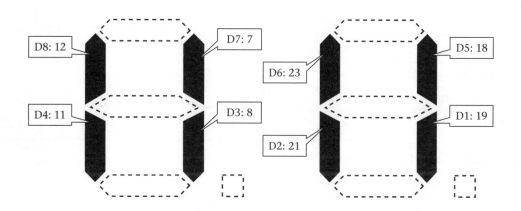

FLEX Display: 8 Individual LEDs

Figure E-2 Altera UP Board FLEX Digital Display Pin Numbers

b. You need to turn off all of the other unused LEDs in order to block the MAX+PLUS II software from randomly routing logic signals through these pins and turning them on. Having unused LEDs inadvertently turn on is a visual distraction. Active low LEDs can be turned off with a connection to logic 1. Connect OUTPUT symbols for the unused LEDs to a logic1 symbol (Vcc). Assign pin numbers to remaining OUTPUT symbols to turn off the unused LEDs.

15$_{\text{FLEX}}$. Use the compiler to generate the output files for simulation and IC programming. Some PCs need the "Quartus Fitter" to be turned *off* to be able to compile without generating a fatal error. Other PCs will work with "Quartus Fitter" *on*. The "Quartus Fitter" is *on* by default. To turn *off* the "Quartus Fitter" follow these steps:

From the MAX+PLUS II menu select **Compiler**. The "Compiler" window will open and allow you to select the **Processing** menu. From the **Processing** menu select

Fitter Settings.... This will open the "Fitter Settings" window. From the "Fitter Settings" window uncheck **Use Quartus Fitter for FLEX10K and ACEX1k Devices** and click "OK."

The "Quartus fitter" settings will be stored with the project files. You need only turn it *off* the very first time you compile. You are now ready to compile the project. From the **File** menu select **Project Save & Compile**.

16_{FLEX}. Power the UP board, connect the Byte Blaster cable, and plug in the power pack.

17_{FLEX}. Program the FLEX IC.

    a. From the **MAX+PLUS II** menu select **Programmer**.

    b. From the menu that runs across the very top of the screen, click on **JTAG**. Make sure there is a check mark beside the Multi-Device JTAG Chain menu item.

    c. From the menu at the top of the screen, click on **JTAG**. Choose **Multi-Device JTAG Chain Setup**. The Multi-Device JTAG Chain Setup window will open. Refer to Figure E-3.

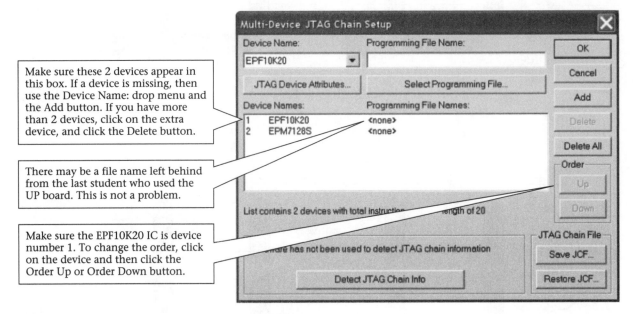

Make sure these 2 devices appear in this box. If a device is missing, then use the Device Name: drop menu and the Add button. If you have more than 2 devices, click on the extra device, and click the Delete button.

There may be a file name left behind from the last student who used the UP board. This is not a problem.

Make sure the EPF10K20 IC is device number 1. To change the order, click on the device and then click the Order Up or Order Down button.

Figure E-3 The Multi-Device JTAG Chain Setup Window

To finish the programming procedure, follow these steps:

1. Click on device 1 **EPF10K20** (or EPF10K70 for UP-2 board).
2. Click the **Select Programming File** button. A window will open that will allow you to select the **lab.sof**. Click the "OK" button.
3. When you return to the Multi-Device JTAG Chain Setup window, click the **Change** button and then click the "OK" button.
4. When you return to the Programmer window, click the **Configure** button and the design will be transferred to the FLEX IC.

Appendix

Altera Simulator Guide

The information presented in this appendix will teach you how to use the Altera **MAX+PLUS II simulator**. A **simulator** is a software package that allows you to test a design before it is programmed into a CPLD. A simulation can help find design errors. The steps required to use the simulator will be applied to the vending machine system of Lab 1. This appendix can also be used as a guide to run simulations for the other simulation lab exercises described in this book.

1. Begin by inserting the disk that has the vending machine files for Lab 1 into the computer. Run the MAX +PLUS II program. Open the file **lab1.gdf** and **Save & Compile** this project. This step may not be necessary; however, it ensures that the MAX+PLUS II simulator is ready to use the current project.

   You need to create a **Simulation Channel File (SCF)**. The Altera simulator uses a program called **Waveform Editor** to create and view the **SCF**. Step 2 will show you how to create an SCF file.

2. Click the **File** menu and select **New....** From the "New" window select **Waveform Editor file** and then click the "OK" button. You will see the "Waveform Editor" window open. Refer to Figure F-1.

   Click the **File** menu, select **Project**, and then **Set Project to Current File**. This will open the "Save As" window. Use the name **lab1.scf** to save the file.

   NOTE: Always use the same file name for both the gdf and the scf files.

   You need to use the "Waveform Editor" to enter waveforms that represent the inputs of your design. Once this is done, the simulator will be able to generate an output waveform to represent the response of your system. You must then analyze the output waveform to figure out if your design works. The next step will use the "Waveform Editor" to add a position marker (called a **node**) for the input and output waveforms. Once the nodes are defined, you will be able to proceed with the definition of the input waveforms and run the simulation

3. With the selection pointer (pointer you can move with the mouse) double-click in the open area just below the label "Name:" You will see the "Insert Node" window. Click the **List** button. Refer to Figure F-2.

   NOTE: If the "List" button cannot be pressed, close the window, "Save and Compile" the project, and then retry this step.

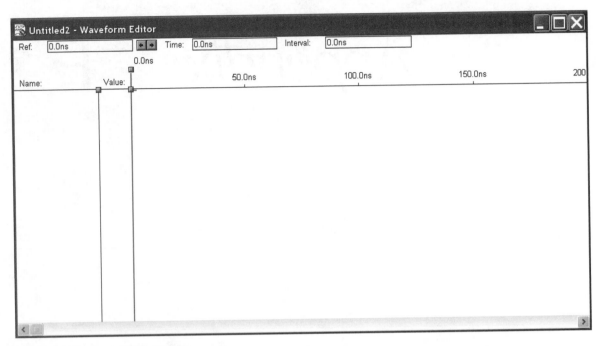

Figure F-1 Waveform Editor Window

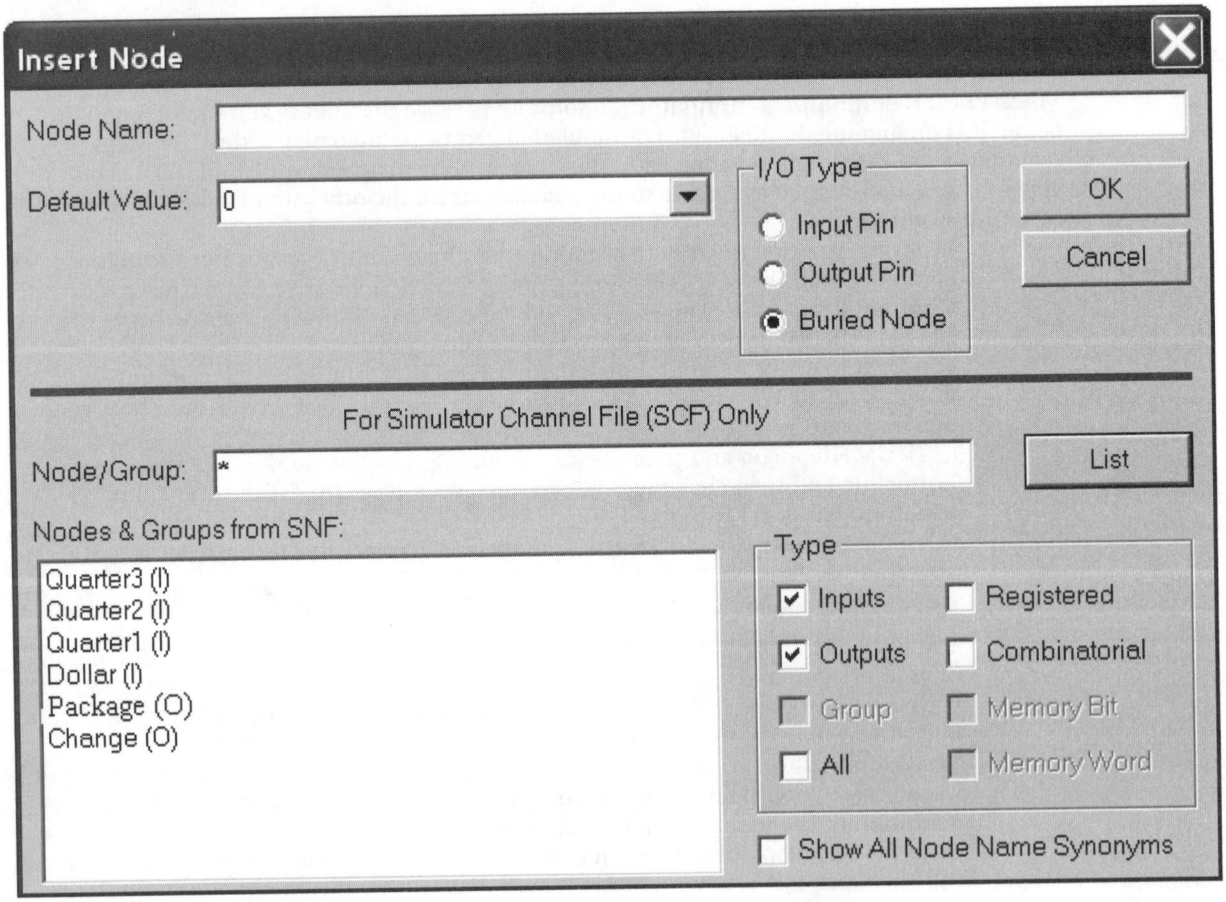

Figure F-2 Insert Node Window

a. Click on node **Quarter1 (I)** and then click on the "OK" button. The "Insert Node" window will close and the "Waveform Editor" window will show the "Quarter1" input. Refer to Figure F-3.

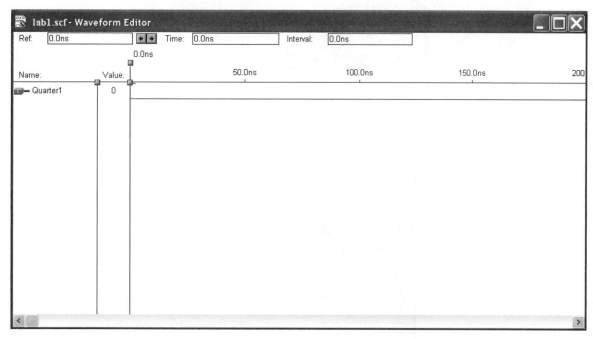

Figure F-3 Waveform Editor Window with Input Quarter1

b. Repeat this procedure to enter nodes for "Quarter2," "Quater3," "Dollar," "Package," and "Change." When you are done the "Waveform Editor" window should resemble Figure F-4.

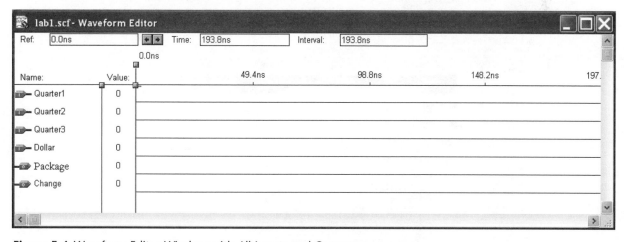

Figure F-4 Waveform Editor Window with All Inputs and Outputs

Before the input waveforms can be entered, you must set the "End Time" and the "Grid Size." The "End Time" sets the total length of time of the simulation. The "Grid Size" sets the size of the smallest increment of time used by the simulation.

4. From the **File** menu select **End Time** .... This will open the "End Time" window. Refer to Figure F-5.

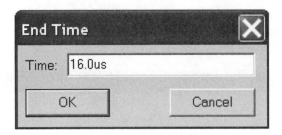

Figure F-5 End Time Window

Set the **End Time** to **16.0us** and press the "OK" button. To adjust the "Waveform Editor" window to view the new "16us" simulation time interval, do the following:

a. From the **View** menu select **Fit in Window**.

b. From the **Options** menu select **Grid Size ....** This will open the "Grid Size" window. Refer to Figure F-6. Set the **Grid Size** to **1.0us** and press the "OK" button.

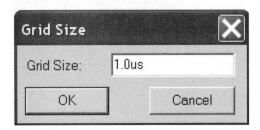

Figure F-6 Grid Size Window

With the nodes and time settings defined, you can now enter the input waveforms. The simulator uses the input waveforms to represent the input data from a truth table. The waveforms represent the 16 truth table entries from 0000 to 1111 (0 to 15).

5. Place the selection pointer over the node named "Quarter1." Right click then select **Overwrite** then **Clock ....** Refer to Figure F-7. The "Overwrite clock" window will open. Refer to Figure F-8. Press the "OK" button and the waveform for node "Quarter1" will appear in the "Waveform Editor" window. Refer to Figure F-9.

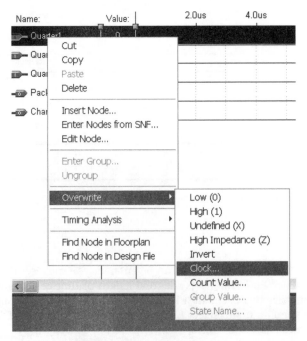

Figure F-7 Right Click over the Node Quarter1

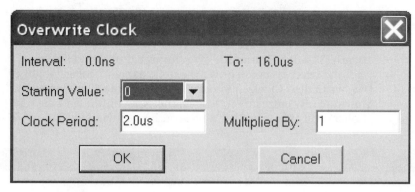

Figure F-8 Overwrite Clock Window

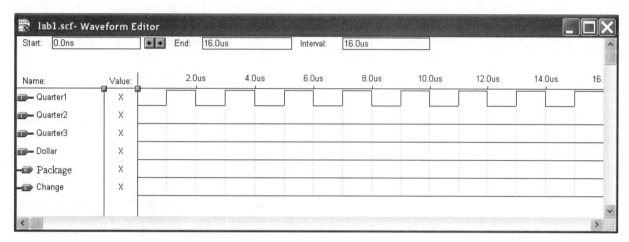

Figure F-9 Waveform Editor Window with Quarter1 Waveform

a. Place the selection pointer over the node named "Quarter2." Right click then select **Overwrite** then **Clock ...** to open the "Overwrite Clock" window. Refer to Figure F-7. Change the **Multiplied By** entry box to **2**. Press the "OK" button and the waveform for node "Quarter2" will appear in the "Waveform Editor" window. Refer to Figure F-10.

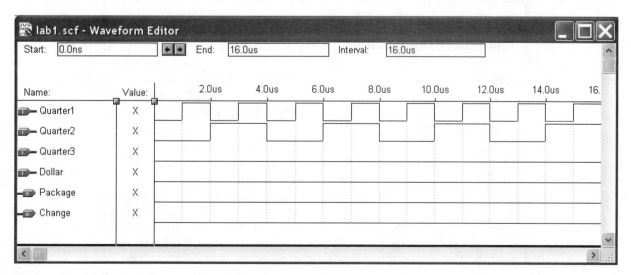

Figure F-10 Waveform Editor Window with Quarter1 and Quarter2 Waveform

b. Place the selection pointer over the node named "Quarter3." Right click then select **Overwrite** then **Clock ...** to open the "Overwrite Clock" window. Change the **Multiplied By** entry box to **4**. Press the "OK" button and the waveform for node "Quarter3" will appear in the "Waveform Editor" window.

c. Place the selection pointer over the node named "Dollar." Right click then select **Overwrite** then **Clock ...** to open the "Overwrite Clock" window. Change the **Multiplied By** entry box to **8**. Press the "OK" button and the waveform for node "Dollar" will appear in the "Waveform Editor" window. Refer to Figure F-11.

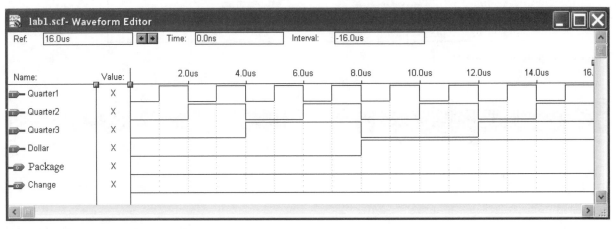

Figure F-11 Waveform Editor Window with All Input Waveforms

The simulation input waveforms have been entered. The waveforms will simulate all 16 truth table entries over a 16-microsecond time interval. The left side of the waveform diagram represents truth table condition "0000" and the right side "1111." It is time to run the simulation and analyze the output results.

6. From the **MAX+PLUS II** menu select **Simulator**. The simulator window will open. Refer to Figure F-12. Press the **Start** button and then the **Open SCF** button. The "Waveform Editor" window will show the results of the simulation. Refer to Figure F-13. The simulation shows the "Package" output = "0" when three quarters are entered (logic 1) and the "Change" output = "0" when the dollar is entered. This is the correct response. The system is designed to turn on active low LEDs when the money is inserted.

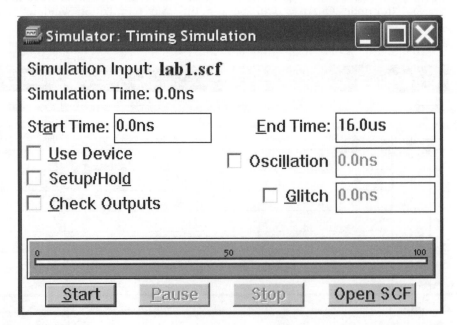

Figure F-12 Simulator Window

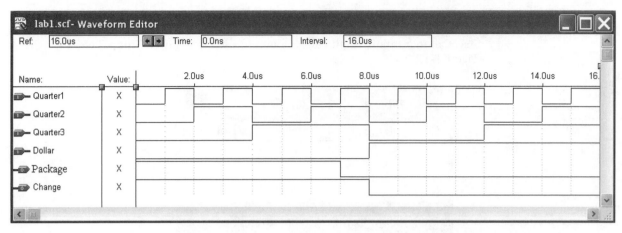

Figure F-13 Simulation Results

NOTE: Simulation Glitches:

*The FLEX IC can cause the simulator to generate **glitches**. "Glitches" are tiny vertical spikes that represent a logic level transient. To produce a clean simulation without "glitches," you can try the following:*

- *From the **MAX+PLUS II** menu select **Compiler**. From the **Processing** menu select **Functional SNF Extractor**. This will place a check mark beside this menu item.*

- *From the **File** menu select **Project** and then select **Save, Compile & Simulate**. This will create a new set of project files and open the "Simulator Window."*

- *Press the **Start** button and then the **Open SCF** button. The "Waveform Editor" window should show the results of a glitch-free simulation. Refer to Figure F-13.*

- *Remember to remove the check mark from the menu item "Functional SNF Extractor" when you have completed the simulation work.*

The input waveforms can be grouped into a "single node." The new "single node" will list each truth table entry numerically. The feature, however, requires that you place the waveforms in order from MSB to LSB. MSB is input "Dollar" and LSB is input "Quarter1." Step 7 will show you how to order and group the inputs.

7. Change the order of the input waveforms. Place the selection pointer over the input symbol for the node named "Dollar." Be sure the selection pointer is *not* placed over the actual text "Dollar." Click and drag the "Dollar" node to the top of the waveform diagram. Repeat this process in order to reorder the other waveforms from MSB to LSB. When you are done, the "Waveform Editor" will resemble Figure F-14.

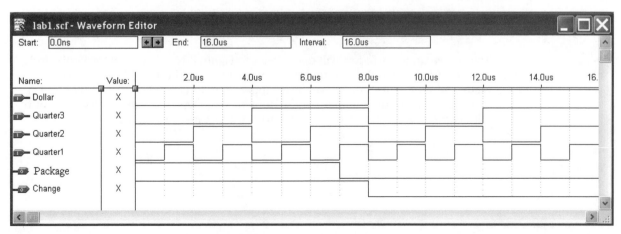

Figure F-14 Waveform editor with Input waveforms reordered from MSB to LSB

- Place the selection pointer over the node named "Dollar." Be sure the selection pointer is placed over the actual text "Dollar." Click and drag the pointer down to select all four inputs. Right click the mouse and select **Enter Group**. Refer to Figure F-15.

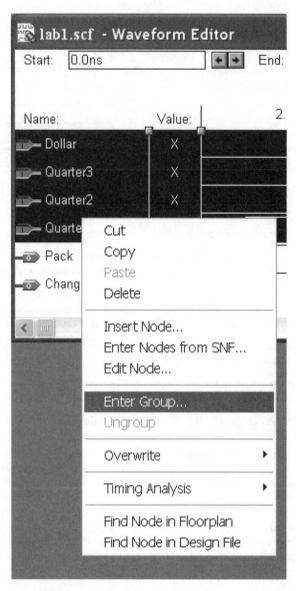

Figure F-15 Click, drag, and select all four inputs and then right click

The "Enter Group" window will open. Refer to Figure F-16. Select **BIN** for **Radix** and enter **money[3..0]** as the "Group Name." Click the "OK" button. The "Waveform Editor" window will show the grouped inputs. Refer to Figure F-17. The 4-bit number inside each box of the combined node "money[3..0]" represents the binary value of the 4 input waveforms. It also corresponds with the entries in the truth table.

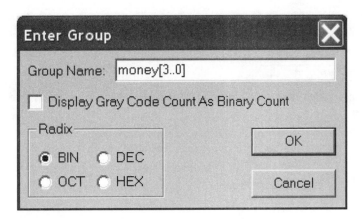

Figure F-16 Enter Group Window

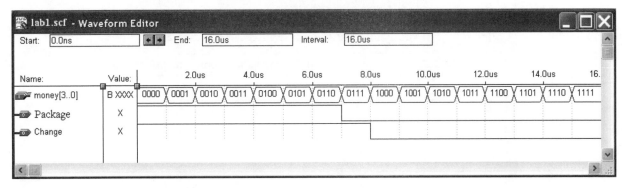

Figure F-17 Waveform Editor with Inputs Grouped

8. This step identifies some useful simulation buttons. To test these buttons, you should ungroup the node "money [3..0]." Place the selection pointer over the node named "money [3..0]." Right click the mouse and select **Ungroup**. The Waveform Editor window will show the ungrouped inputs and revert back to Figure F-14. Explore the useful simulation buttons for yourself. Refer to Figure F-18.

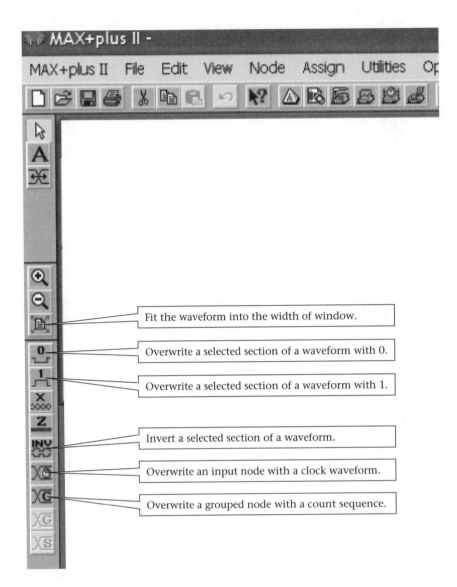

Figure F-18 Simulator Buttons

Index